PAPUA NEW GUINEA

Science

Grade 6

Student Book

Kenneth Rouse

OXFORD

Level 8, 737 Bourke Street, Docklands, Victoria 3008, Australia

Oxford University Press is a department of the University of Oxford. It furthers the University's objective of excellence in research, scholarship, and education by publishing worldwide in

Oxford New York

Auckland Cape Town Dar es Salaam Hong Kong Karachi
Kuala Lumpur Madrid Melbourne Mexico City Nairobi
New Delhi Shanghai Taipei Toronto

With offices in

Argentina Austria Brazil Chile Czech Republic France Greece
Guatemala Hungary Italy Japan Poland Portugal Singapore
South Korea Switzerland Thailand Turkey Ukraine Vietnam

First published 2007
Reprinted 2007, 2008 (twice), 2009 (twice), 2010, 2013, 2014 (twice), 2015, 2017, 2019, 2022, 2024

978 0 19 555114 3

Typeset by Pier Vido
Illustrated by Uramina & Nelson Ltd
Printed in China by Golden Cup Printing Co. Ltd

Oxford University Press Australia & New Zealand is committed to sourcing paper responsibly.

Contents

1

Working scientifically

Chapter summary

In this chapter you will have an opportunity to:

- learn how to ask questions that will guide your observations in order to understand the world around you
- carry out simple tests and describe your observations
- identify groupings of things and patterns of information that will allow you to make conclusions
- learn how to improve the way that you find out information
- find out how science is used in everyday life.

Syllabus references

Strand: Working scientifically

Sub-strand: Working scientifically

Outcomes: **6.1.1** Investigate the immediate environment and using scientific methods, organise their experiences and communicate their ideas

Key facts

- People try to understand their surroundings and often try to find out how and why things happen.
- People in PNG have much traditional knowledge about the living things that are found in their surroundings, the materials that are available in the local area and how to use them.
- Science is also a way of finding out and understanding the world around us.
- One way of finding out about our surroundings is to make observations and to ask questions.
- Scientists measure things in order to be able to compare them and understand them better.
- Scientists describe their observations and carry out tests to find out more about their ideas.
- Scientists try to look for groupings and patterns in the information that they collect and also to make conclusions about the things that they understand.
- We can use our knowledge from our observations and description of patterns to make predictions about what will happen in the future.
- Scientists try to organise their knowledge and ideas so that they can share them with other people.
- The way that scientists try to find out or investigate the world around us and communicate their ideas is called the scientific method.
- We use science every day and this can help to improve our lives.

What is science? PD SS

For a long time people all over the world have tried to understand the things that they observed happening around them and to make use of the things they found in the surroundings. For example, people in Papua New Guinea have developed a lot of knowledge and understanding about how to live in their local area. Much of this knowledge has been learned from experience and has been passed on from parents, uncles and aunties to their children. This type of knowledge and these skills are sometimes called **traditional knowledge**.

Although there are differences between the highlands and the islands, between coastal and inland areas, all people have learned how to grow food and catch fish and other animals to eat. They have also learned how to make fire, how to cook food and how to smoke meat and fish so that it could be kept and used later. They learned to build houses and use materials that are found in the local area. For example, many people learned how to make pottery and this was traded between people from different places. People have learned that some plants can be used as medicines to make people better when they get sick. People also understood the cycle of seasons and how this can affect the food that they grow, and when it is good or not so good to travel to other places. People also knew how to use the

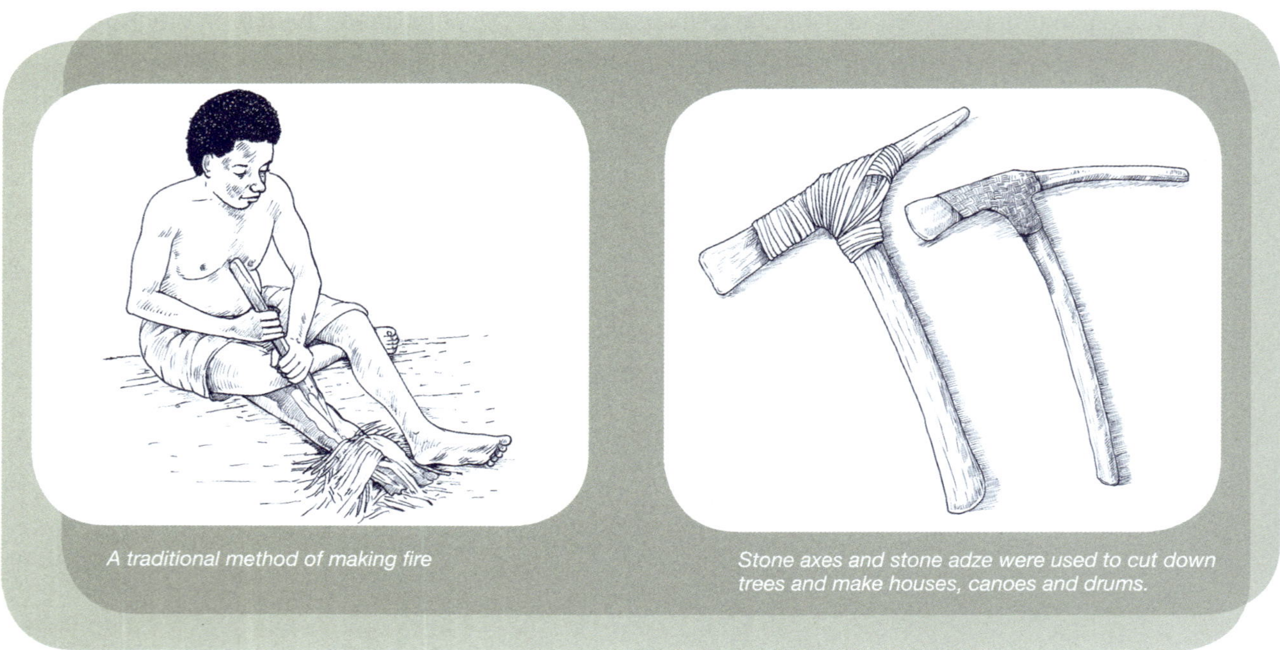

A traditional method of making fire

Stone axes and stone adze were used to cut down trees and make houses, canoes and drums.

People used kundu drums for traditional dancing.

moon and the stars to decide when to go fishing or hunting and when to plant their food crops.

People learned how to make tools like stone axes and stone adzes and these were used to cut trees and to make things out of wood, like the kundu and the garamut. They used a type of very hard, black shiny stone called **obsidian** to make a sharp cutting edge. People living by the sea and close to rivers and lakes also learned how to make canoes and some coastal people travelled long distances along the coast or between the islands for traditional trading. They understood how to find their way or how to navigate, using the direction of the wind, the current and the stars.

Making canoes, being able to navigate and knowing when the season changes are all important for traditional trade in Papua New Guinea.

• Provincial capital
EHP Eastern Highlands
SHP Southern Highlands
WHP Western Highlands
NCD National Capital District

• Pots
○ Axes and knives
+ Salt springs

0 50 100 150 200 km

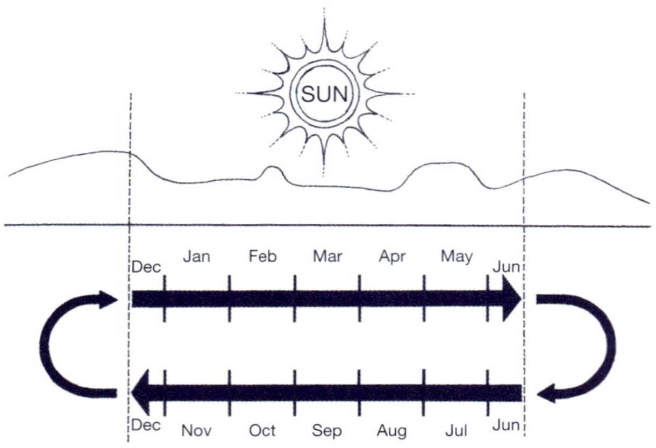

A horizon calendar

They used a traditional calendar such as the **horizon calendar** to predict when the seasons would change. This knowledge was used to decide the right time for gardening, for arranging special ceremonies and for organising trading expeditions.

When people use and understand the things around them, they notice patterns and then put the things into **groups** depending on their properties or characteristics. Putting things into groups according to their similarities and differences is called **classification**. For example, the Wola people of the Southern Highlands group together all large wallabies and call them *sab*. They group together all woody trees and call them *iysh*. They can also tell the difference between many different kinds of sweet potato and taro. The Karam people who live in the mountains of Madang Province say that the cassowary is not a bird because it cannot fly. Other people say that the cassowary is a bird because it lays eggs and has feathers and wings like other birds, even though it cannot fly. Within each society and language group the system of classification is well known but different societies use different methods of classification.

Many people in Papua New Guinea also developed their own system of **counting** and some developed ways to **measure** the size of things so that they could be compared more easily. And, of course, everyone developed their own language so that they could communicate with each other.

Counting in the Mengen language (from Pomio in East New Britain)

one	kena	six	lima ba kena
two	lua	seven	lima ba lua
three	mologi	eight	lima ba mologi
four	tugulu	nine	lima ba tugulu
five	lima	ten	tanulel

These examples show the way that people have tried to use the things available in their surroundings in order to live, and also to understand the world and the relationships between different things.

People in other countries have also been learning about how to live in their surroundings and trying to understand the things that they could see and hear and touch. One of the main ways that people do this is to ask questions about how and why things happen. For example, they may ask questions like these:

- Where do different materials like metals and plastic come from and how are they made?
- How can we use different materials to help us?
- How are machines made and how do they work?
- What are the similarities and differences between living things?
- Why do we have different seasons?
- Why do the sun, moon and stars rise and set in the sky?

Science is about the study of nature and how the world works. Science includes finding out about naturally occurring things such as rocks, mountains, the moon and the stars, as well as things made by people such as tools, machines, radios and computers. Science includes understanding living things like plants and animals and the reasons why we get sick and why we get better.

However, science is not just about facts and knowledge. Science is also:

- a way of doing things
- a way of looking at things
- a way of thinking about things.

The work of scientists

People who work in science are called **scientists**. Scientists are always asking questions about the world around them and then trying to find answers to their questions. Scientists have found the answers to many questions, but there are many questions yet to be answered. People also keep finding new questions.

There are many different kinds of scientists and so science can be divided into a number of different branches.

Branches of science

Branch of science	What is this branch interested in?
Archaeology	The things that people have left behind like tools and pottery that help to find out about the past and the way that people lived
Astronomy	The planets, sun, moon, stars and the universe
Biology	Living things like plants and animals
Chemistry	Substances like chemicals and medicines and the way that they behave
Ecology	How plants and animals live and interact in the environment
Geology	Rocks and the earth; looking for oil and gas
Health	The reasons why we get sick and how we can get better
Meteorology	The weather, which is caused by what is happening in the atmosphere
Physics	Forces and movement, energy and matter, tools and machines

For you to try

1 Copy and complete the following table about traditional knowledge for the people living in your area, or in an area that you know well. To help you, one example has already been done. A

Traditional knowledge	What do people know or what do they do?	How is this useful?
Making fire	Choose two special kinds of wood and rub one very hard against the other	Fire is needed to cook food, to keep warm, to make pottery
Making pottery		

2 Find out about a system of grouping or classification in your area or an area that you know well. Make lists to show this information. For each list give the name of the group and some examples from the group. What similarities and differences do people use to classify each group?

Share your information with others and discuss the differences.

3 People in Papua New Guinea have scientific knowledge but did not use  the word **science** or the different branches of science to describe their traditional knowledge and skills. Using the examples from your answer to **Activity 1**, copy and complete the following table to show how the knowledge or skill of people in Papua New Guinea can be put into one of the branches of science. Remember that some knowledge or skills may belong to more than one branch of science. To help you, one example has already been done.

Traditional knowledge or skill	Branch of science
Growing food in the garden	Biology and ecology

Using your senses for making observations

We all use our senses to find out what is going on around us. We gather information which helps us to understand what is happening. For example:

- We use our eyes to see the shape and colour of things. This is called the sense of **sight**.
- We use our ears to hear sounds and noises. This is called the sense of **hearing**.
- Our sense of **taste** helps us to know if food is sweet, sour, salty or bitter.
- Our sense of **smell** can tell us if something is rotten or if something is burning.
- Our sense of **touch** can tell us if things are rough or smooth, hot or cold, wet or dry.

We can also feel the heaviness of things. When we lift or carry things we know how heavy they are. For example, a bag of cement usually weighs 50 kg and so is twice as heavy as a 25 kg bag of rice although they are about the same size. A carton of corned beef is much heavier than a carton of Twisties although there is not much difference in the size.

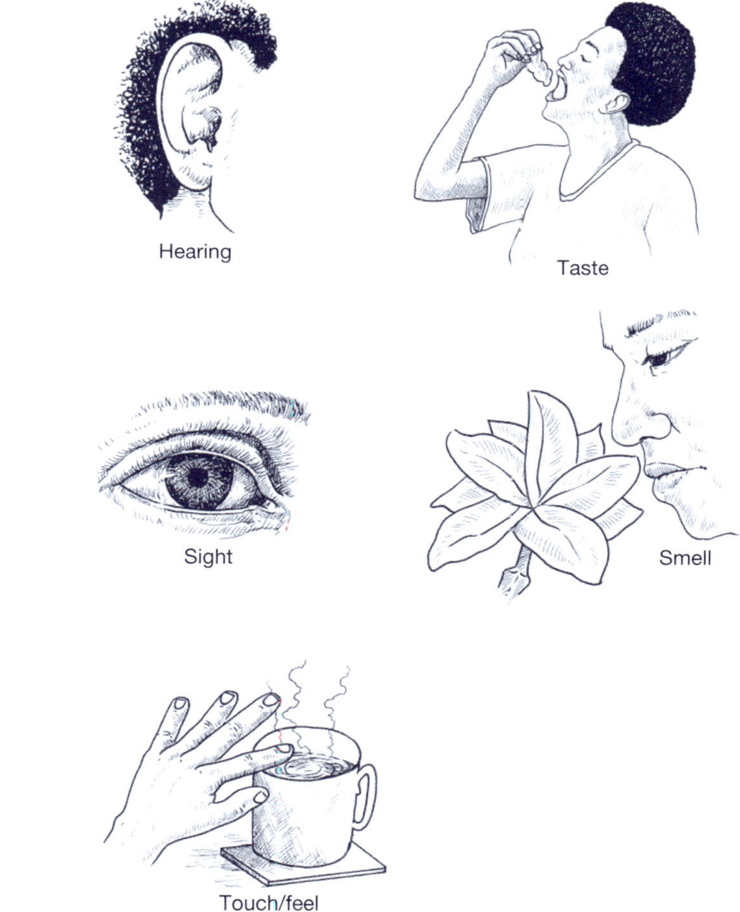

The five senses

More than observations

Gathering information or making observations is something that we all need to do every day. But after we collect information we usually do something with the information—we might try to understand what it means and make some **conclusions**, and might decide that we have to do something about it. For example, if you fall asleep in the afternoon and wake up because you can feel the house shaking, then you might think that there is an earthquake and run outside because you feel scared. If you smell that something is burning but you cannot see anything burning you might think that somebody in your family has left a pot of food on the stove or on the fire and you might go to find out what has happened.

When you conclude something, you are also making an **inference**. An inference is also a **possible explanation** of your observations. Of course, when we make an inference we can sometimes be wrong. We need to make more observations in order to find out if our inference is correct. For example, if you run outside when you feel the house shaking and find that your friends were making the house shake, you know that they were tricking you and it was not an earthquake. If you find that your food is not burning in your kitchen then there must be some other explanation for the smell. Maybe the smell came from the house next door or further away and was carried by the wind.

Observation is also an important skill that we use in science. After making observations we try to understand our observations or to find some explanation. For example, we might make some conclusions or inferences.

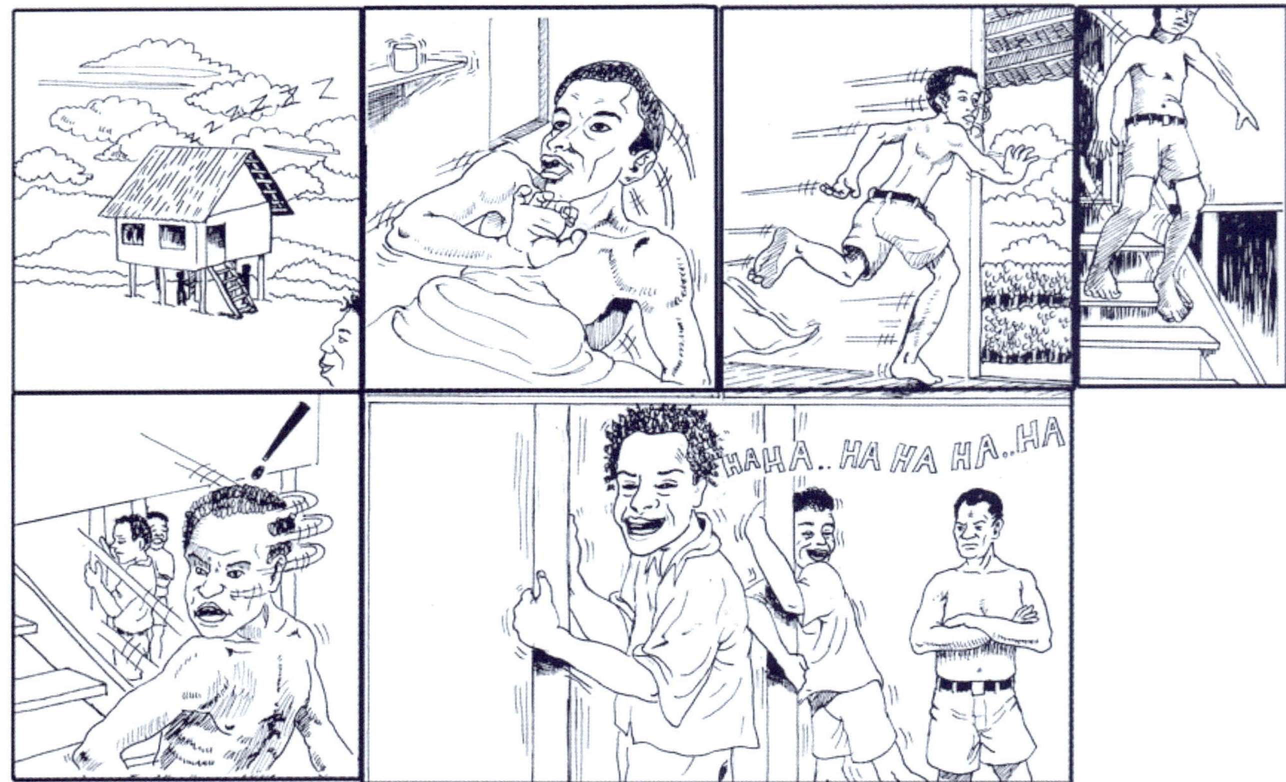

An earthquake, or friends playing a trick?

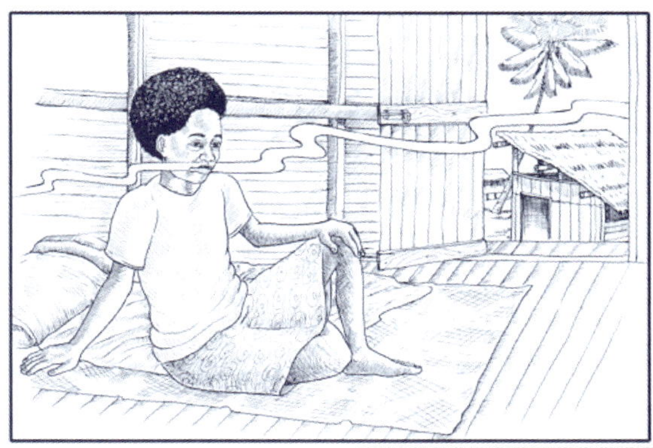

Is somebody's food burning?

For you to try

1 Name the five senses used in making observations.

2 What observations can you make sitting in your classroom without moving from your place? Remember to use all your senses and try to give some explanation. Copy and complete the following table to help you. One example has been done for you. Are there any senses that you have not used?

Observation	Senses used	Possible explanation — what does this mean?
Sound of students laughing in the classroom next door	Hearing	The teacher made a joke.

3 **Leader of the orchestra**: a game of observation

- The whole class sits in a big circle.
- One person is chosen to be the observer. This person must be sent away for a minute or two while the leader of the orchestra is chosen.
- The leader of the orchestra starts to pretend that he or she is playing an instrument and everyone else follows the leader. For example, the leader might move her arms and body as if she were playing a drum.
- The observer comes back, stands in the middle of the circle and looks at the orchestra.
- The leader then changes to play another imaginary instrument and again everyone else in the circle follows while the observer must try to guess who is the leader.
- To be fair, the leader must keep changing instruments, so he or she must have different ideas ready (drum, guitar, trumpet, keyboard or piano, ukulele, bamboo band, organ, bagpipes, etc)
- The members of the orchestra must not all look directly at the leader or they will give the game away.
- The game continues until the person in the middle finds the leader of the orchestra. Two new people can then be chosen.
- After a certain number of rounds each person who was an observer can explain how they tried to discover who the leader was.

Things that we do in science

Babies and small children like to touch, taste, squeeze, throw and drop things. In this way they learn the rules that make things behave in the way they do. When small children play, they are doing little experiments and finding out about the world around them. For example, children find out that when they drop something it always falls downwards, although they don't understand why this happens.

Scientists also carry out experiments to find out about the world around us but they follow a number of steps called **scientific processes**. When we do the following, we are using the processes used in science.

- Ask questions in order to find the answer.
- Make observations—using all our senses.
- Carry out experiments or investigations so that we test our ideas.
- Make measurements using different equipment or instruments.
- Collect information from our observations and measurements.
- Record information so that we can organise it, understand it and share it with others.
- Put things into groups or classify them depending on their characteristics.
- Solve problems so that we have a better understanding or make our lives easier.
- Make predictions about what will happen next time or in the future.

The scientific method

Over many years scientists have developed a particular way of working and solving problems which is called the **scientific method**. The scientific method usually follows some or all of the following steps:

1 Asking a question that guides our observations
2 Suggesting a possible answer
3 Testing the ideas by doing an experiment
4 Recording the results of the experiment
5 Checking if the results support the possible answer that was suggested before
6 Thinking again and carrying out more experiments if needed
7 Making final conclusions.

When we use the scientific method we must be careful in our observations and make good judgments. We must collect accurate information and have good reasons for the conclusions that we make.

Using the scientific method

The following simple example shows the way that the scientific method can be used to answer a question or solve a problem.

The question or problem

Can we lift a heavy object using our breath?

Possible answers

We can lift a heavy object using our breath, *or* we cannot lift a heavy object using our breath.

Carry out an experiment

- Collect a large plastic shopping bag and lay it flat on a table or desk so that the open end hangs over the edge. The bag must not have any holes.
- Put a large book on the bag.
- Gather the open end of the bag together to make a mouthpiece and blow steadily into the bag.
- Observe what happens
- Ask a student to sit on the book and repeat the experiment.
- Again observe what happens.

Record your results

The book was lifted by the air inside the bag. When a person sat on the book the person was also lifted by the air inside the bag.

Conclusion

We can lift a heavy object using our breath.

For you to try

1 Do the following investigation: **Can we lift a heavy object using our breath?** SR
   - Collect a large plastic shopping bag and a large book with a hard cover.
   - Carry out the experiment in the box above to find your own result.
   - Describe what you did and what you found out.

2 Are you observant? How much notice do you take of the things that surround you? Without checking first, write down the answers to the following questions:

   a Which side of the classroom gets the most sunlight in the morning?

   b What part of the school grounds gets the most shade at lunchtime and in the afternoon?

   c Which is longer, your big toe or the one next to it?

   d Which is closest in time, the last school holiday or the next school holiday?

   e If there is no cloud, will you be able to see the moon in the sky tonight? If so, what shape will it be?

3 Your teacher will put about ten objects on a tray covered with a cloth. The cloth will be removed for one minute for you to observe the objects. After one minute the objects will be covered again. Write down as many objects as you can remember.

4 Look carefully at the two diagrams below and list as many differences as you can.

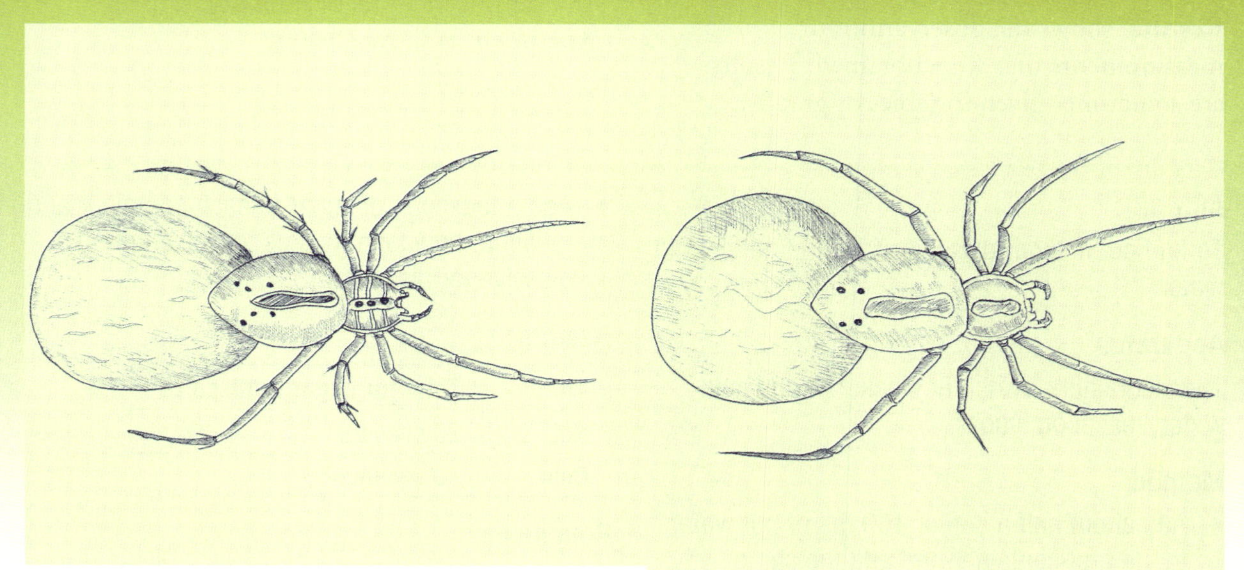

Ways of working scientifically

When you work scientifically, it is important to follow these steps.

- Use a notebook to write down the steps and methods that you use in an experiment.
- Become a good observer. Make a note of what you observe in an experiment. Remember that some observations may not seem important at first but may turn out to be important later.
- Think about what you are doing and what you can work out from the information you collect. Think about all the facts and results before you make a conclusion.
- Always try to find the reason for everything that you observe.

Writing up experiments, investigations or practical reports

L

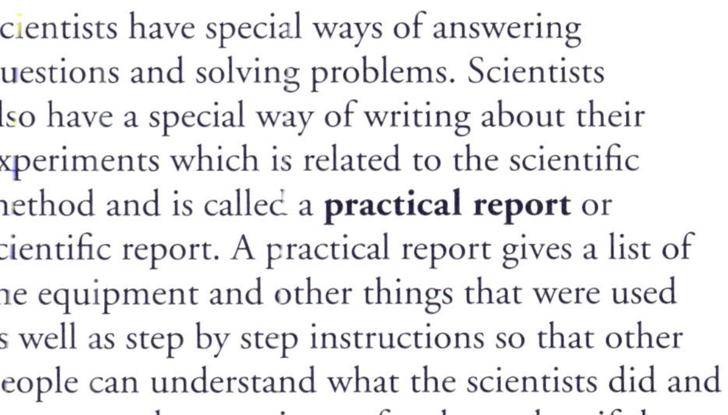

Scientists have special ways of answering questions and solving problems. Scientists also have a special way of writing about their experiments which is related to the scientific method and is called a **practical report** or scientific report. A practical report gives a list of the equipment and other things that were used as well as step by step instructions so that other people can understand what the scientists did and can repeat the experiment for themselves if they want to. Most reports use the following headings.

- **Aim**: states the problem, question or purpose of the experiment.

- **Apparatus**: lists the equipment used in the experiment.
- **Method**: gives the instructions used to carry out the experiment. This sometimes includes diagrams.
- **Results**: shows the observations or measurements that were obtained. Results are sometimes given in tables or graphs. Sometimes a discussion of the results is included.
- **Conclusion**: gives the answer to the aim or problem being investigated.

An example of a practical report from an experiment is shown below.

Aim

To find out if sugar dissolves faster in hot or cold water.

Apparatus

Two containers, watch or clock, hot and cold water, teaspoon, sugar.

Method

- Add about half a cup or 100 mL of hot water to the first container and add the same amount of cold water to the second container.
- Add a teaspoon of sugar to each container and measure how long it takes for the sugar to dissolve (or disappear).

Results

Water	Time for sugar to dissolve
Hot	20 seconds
Cold	40 seconds

Conclusion

Sugar dissolves faster in hot water than in cold water.

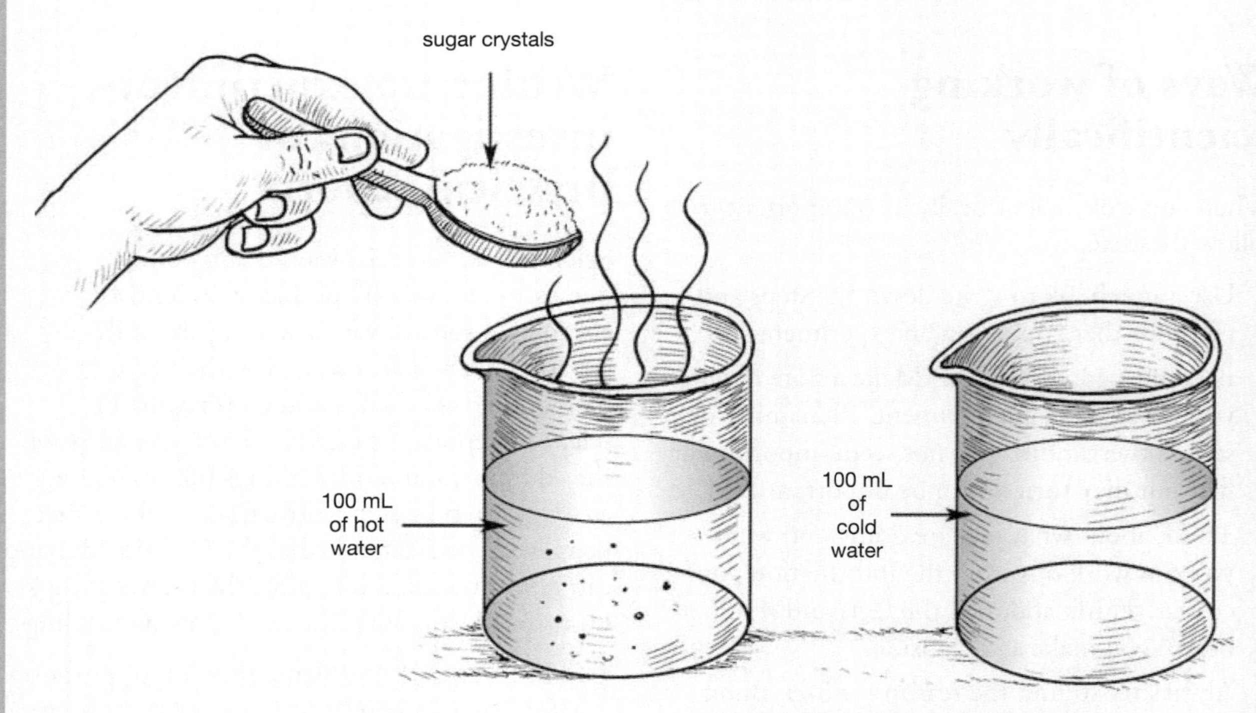

For you to try

1 What senses are used in the experiment above?

2 In the experiment above, why is it important to use the same amount of hot and cold water in each container?

3 Think of an experiment that you could carry out to find out if stirring makes a difference to the speed that sugar dissolves in water. Collect the things you need, carry out your experiment and write up a practical report (in this case you should not change the temperature of the water).

4 How can science be used to improve your life at home or at school?

Summary questions

1 Which of the following are steps or scientific processes that are followed by scientists? Answer **true** or **false**.
   - a Try to find the answer by asking questions.
   - b Use only some of their senses to make observations.
   - c Test their ideas without carrying out experiments or investigations.
   - d Use different equipment or instruments to make measurements.
   - e Use their observations and measurements to collect information.
   - f Record and organise information but do not share it with others.
   - g Classify things or put them into groups depending on their similarities and differences.
   - h Make people's lives harder by solving problems.
   - i Make predictions about what will happen next time or in the future.

2 Science is about facts and knowledge. Which of the following statements are also true about science?
   - A Science is a way of doing things.
   - B Science is a way of looking at things.
   - C Science is a way of thinking about things.
   - D All of the above.

3 Which of the following best describes the order or sequence in which we usually write a scientific report?
   - A Aim, apparatus, method, results, conclusion
   - B Aim, method, apparatus, results, conclusion
   - C Aim, method, conclusion, results
   - D Aim, apparatus, result, conclusion

4 Which of the following statements describe an inference?
   - i An inference means making a conclusion.
   - ii An inference is a possible explanation of results.
   - iii An inference is always correct.

   - A i only
   - B ii only
   - C i and ii only
   - D i, ii and iii

5 The passage below is a summary of the main ideas of this chapter. Copy and complete the passage in your book. Using the words in the list, find the words that are missing. You can use each word only once. (L)

conclusions, day, knowledge, measure, observations, predictions, surroundings, scientific, scientists, share, understanding

People try to understand their ____________ and try to find out how and why things happen. People in PNG have much traditional _____________ about the living things that are found in their surroundings, the materials that are available in the local area and their ways of doing things. Science is also a way of finding out and _______________ the world around us.

One way of finding out about our surroundings is to make _____________ and to ask questions.

Scientists ___________ things in order to be able to compare them and understand them better. _____________ describe their observations and carry out tests to find out more about their ideas. Scientists try to look for groupings and patterns in the information that they collect and also to make ____________ about the things that they understand.

We can use our knowledge from our observations and description of patterns to make ____________ about what will happen in the future. Scientists try to organise their knowledge and ideas so that they can ______ them with other people. The way that scientists try to find out or investigate the world around us and communicate their ideas is called the ____________ method. We use science every ______ and this can help to improve our lives.

2

Living things

Chapter summary

In this chapter you will have an opportunity to:

- find out about the similarities and differences between plants and animals
- find out about the way that plants and animals are made and be able to share this information with others
- find out about the way in which sense organs work and why senses are important
- compare the senses of humans with the senses of other animals
- learn to sort living things into groups using their characteristics
- find out about the relationships between plants and animals in the local area
- find out about the way that animals depend on plants for food.

Syllabus references

Strand: Living things

Sub-strand: Nature of living things

Outcomes:

6.2.1 Identify the basic structure of living things that allow them to function in their environment

6.2.2 Using a diagram find out how energy moves through the living and non-living community

Key facts

- All living things can be placed into groups depending on various characteristics. This is called grouping or classification.
- Similarities and differences between people and things are used to put them in the same group or in different groups.
- The two main groups of living things are animals and plants.
- Animals can be divided into two groups: animals without backbones and animals with backbones.
- Plants can be divided into two groups: flowering plants and non-flowering plants.
- All living things are made of building blocks called cells.
- Plants and animals have different types of cells.
- Cells are grouped together to form tissues. Two or more tissues together form organs. Tissues and organs together form a system that carries out a particular job.
- Many animals have sense organs which allow them to find out about their surroundings and to live successfully.
- Plants and animals live and interact together in the surroundings.
- One important relationship between plants and animals is the food chain. All animals depend on plants for food.

Nature of living things

PD

Papua New Guinea has many living things that are found only in this part of the world. For example, some kinds of kangaroos, wallabies, birds of paradise and butterflies (like the birdwing) are found only in PNG, although Australia also has some of these types of animals as well. Many of these animals are important to the culture of the people of PNG. Feathers and fur, as well as special plants, are often used in traditional dancing, and animals are shown on our coins and stamps. The most important symbol of Papua New Guinea is the bird of paradise, which is part of the national crest. It is shown on all coins and notes. The bird of paradise is also shown on the national flag. The national airline Air Niugini uses the bird of paradise as the company logo as it is a symbol which is recognised all over the world.

The Air Nugini logo

The national crest

The national flag

PNG coins

Special leaves and feathers are used in traditional dancing.

Characteristics of living things

All the things that we find in the world are either living or non-living. Living things belong to a group that has many characteristics that non-living things do not have. In particular, living things carry out a number of activities or processes that allow them to stay alive. For example, living things:

- move
- grow or become bigger—they develop
- react or respond in particular ways to the conditions around them
- reproduce or make copies of themselves
- get energy from food sources during **respiration**
- give off waste products—they excrete.

Living things are made up of **cells**. Cells are the very small building blocks from which all living things are made. Many of the things that we see around us are made of a number of smaller parts or units that are brought together in a particular way in order to achieve a particular purpose. For example, a building can be made from concrete blocks, pieces of timber, and sheets of fibro and corrugated iron. In the same way, a living thing is made of different kinds of cells.

Living things are usually divided into two main groups called **plants** and **animals**. We can tell the difference between plants and animals by looking at three main characteristics:

- their movement
- whether they make their own food or not
- the structure of their cells.

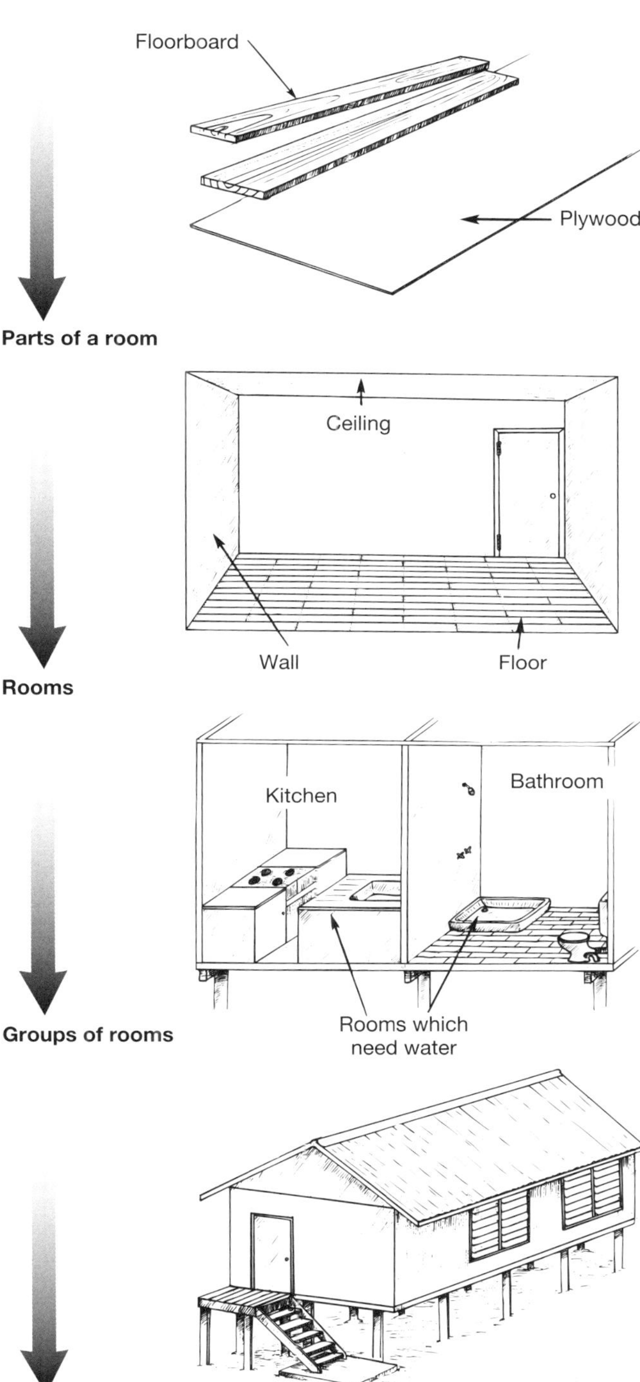

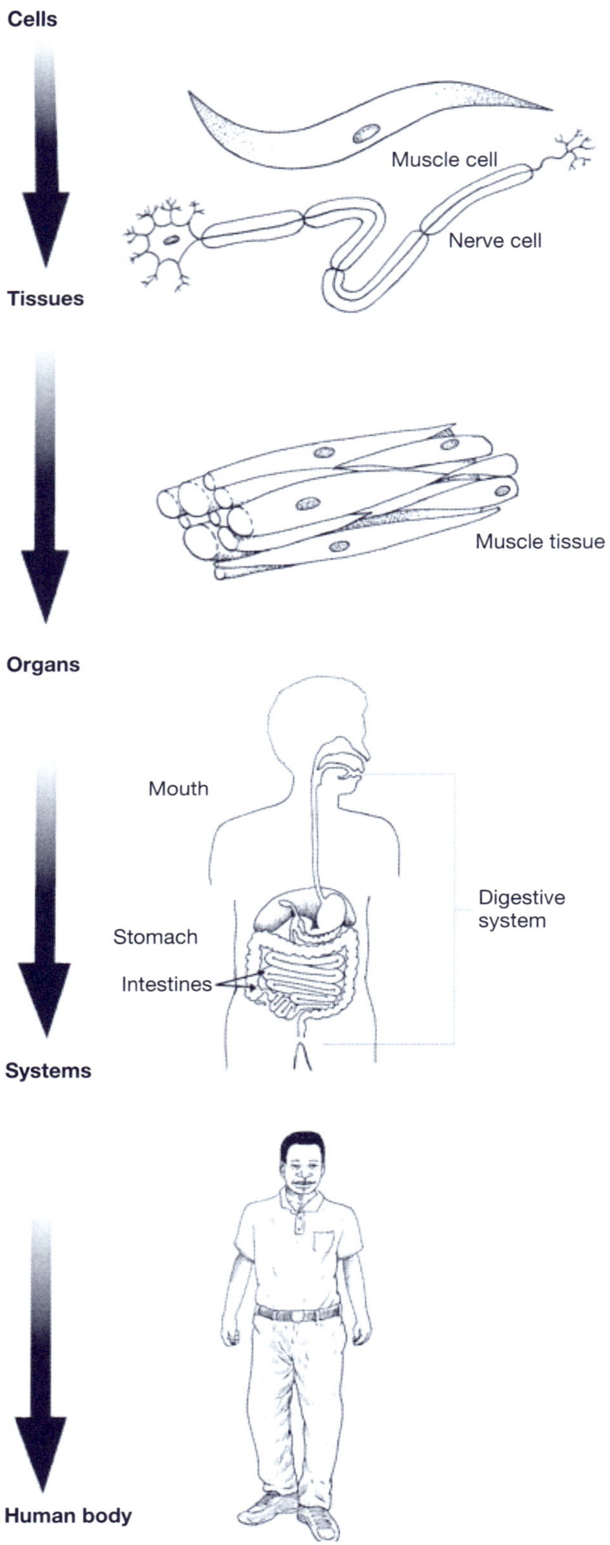

Cells
Muscle cell
Nerve cell
Tissues
Muscle tissue
Organs
Mouth
Stomach
Intestines
Digestive system
Systems
Human body

Living things
How do they move?
Do they make their own food?
What are their cells like?
Plants
Animals

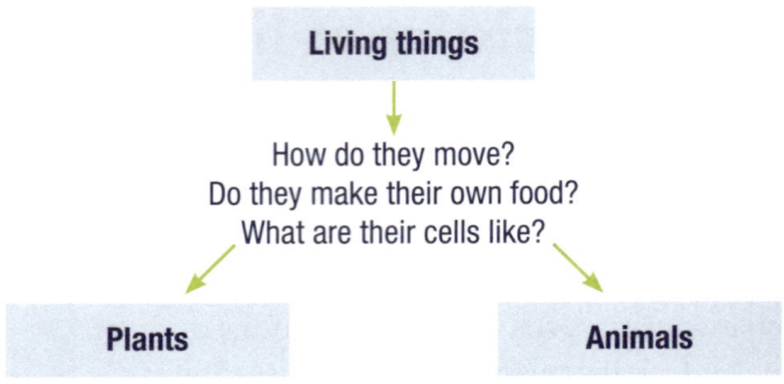

Dividing living things into plants and animals.

Movement

Most plants have roots in the ground or are fixed in one spot and so cannot move from place to place. However, the upper parts of some plants are able to move in particular ways. For example:

- leaves can move with the sun as it moves across the sky so that they receive more light
- in some plants the flower moves to face the sun e.g. the sunflower,
- some flowers and the leaves of some trees close or fold up at night
- the tendrils of climbing plants (e.g. passionfruit and grapes) move in circles until

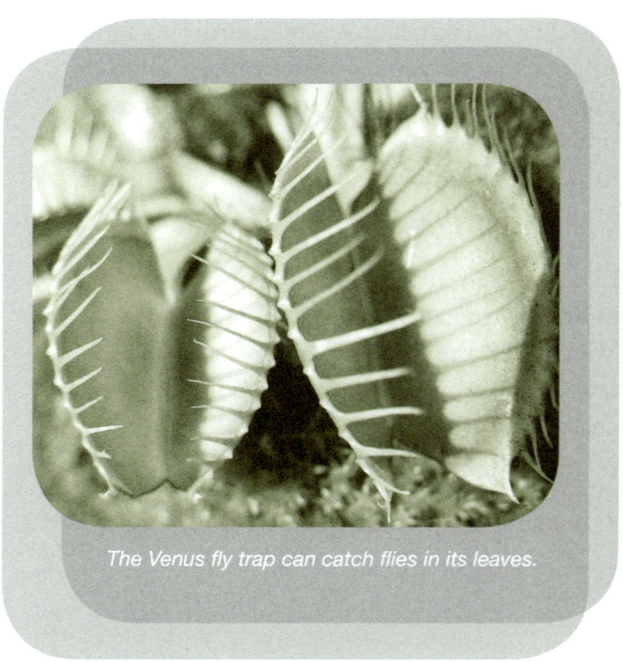

The Venus fly trap can catch flies in its leaves.

they touch something on which they can climb. When they touch something they will wind around it which helps to hold the plant upright.

Most of the movements that plants make are slow, but some plants are able to move more quickly. For example, when you tread on mimosa, or touch it with a stick, the plant will quickly fold up its leaves and then fold the stem. The Venus fly trap can close its special leaves very quickly. It does this to trap the flies that it uses as a source of food.

Most animals are able to move and have various ways of moving from place to place. For example, legs, wings, fins, tails and small hairs are some of the special parts that are used in movement.

Living things move in different ways.

For you to try

Investigation: How do living things respond?

1
- Find some mimosa growing in your area. SR
- Touch it in different places and in different ways. What happens?
- Does it make any difference where and how you touch the plant?
- How long does it take for the leaves of the mimosa to return to their normal position?
- Write up your investigation to explain what you did and what you found out.

2
- Grow some seedlings in a pot and cover them with a box that has a hole in it so that they get light from one side only. SR
- What happens? How long does it take for the plant to respond?
- Write up your investigation to explain what you did and what you found out.

3
- Observe different climbing plants growing in your area.
- What do they climb on? Do they climb in a clockwise direction or counter-clockwise? How high do they climb?
- Write up your investigation to explain what you did and what you found out.

4
- Grow some seedlings in a pot and lay the pot on its side for several days. Make sure the pot does not roll to a different position.
- Observe what happens. Why does the plant do this?
- Write up your investigation to explain what you did and what you found out.

Making or obtaining food

Most plants are able to make their own food. Most plants are green because they contain a special substance called **chlorophyll** which gives them their green colour. However, in some seaweeds and crotons, the green colour is hidden by other colours such as red and brown.

Plants that contain chlorophyll are able to make food by a process called **photosynthesis**. The plant uses energy from the sun and takes a gas called carbon dioxide through the leaves and water through the roots to make food. The process of photosynthesis can be represented by the following equation:

carbon dioxide (from the air) + water (from the soil) = sugar + oxygen

Animals and fungi do not contain chlorophyll and so they cannot make their own food—they must get it from elsewhere. The food that they obtain may come from plants or from animals and may be alive or dead. Some animals such as wasps, spiders and tigers catch other animals for food. Many spiders build special webs that are used to catch moths and other insects for food.

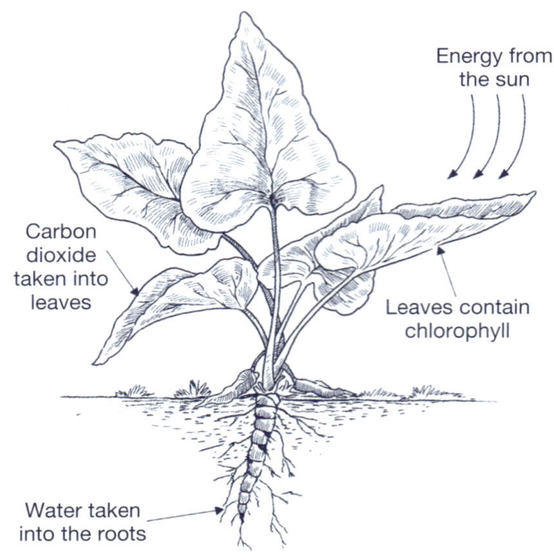

Photosynthesis in green plants

For you to try

Copy and complete the following table to show the different kinds of plants found in your area. You should include plants that live on land and in water. Two examples have been done for you.

Plants that are green	Plants that contain other colours
Grass	Croton

Plant and animal cells

All living things are made of tiny building blocks called **cells**. Cells are so small that they can only be seen with a **microscope**—the instrument that scientists use to make very small things look bigger.

Some living things are so small that they are made from only one cell. However, most animals, including humans, are made of millions of different cells like blood cells, skin cells, muscle cells and nerve cells.

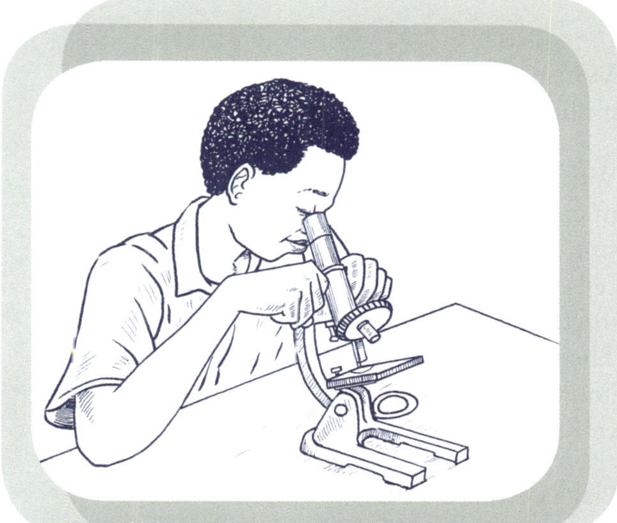

Using a microscope

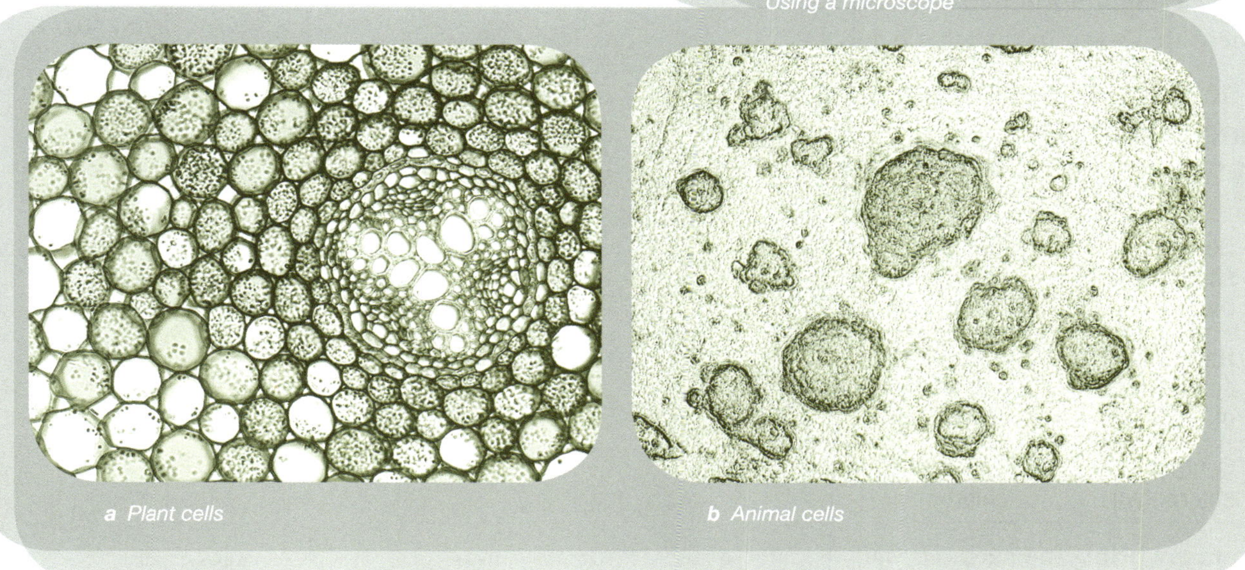

a Plant cells

b Animal cells

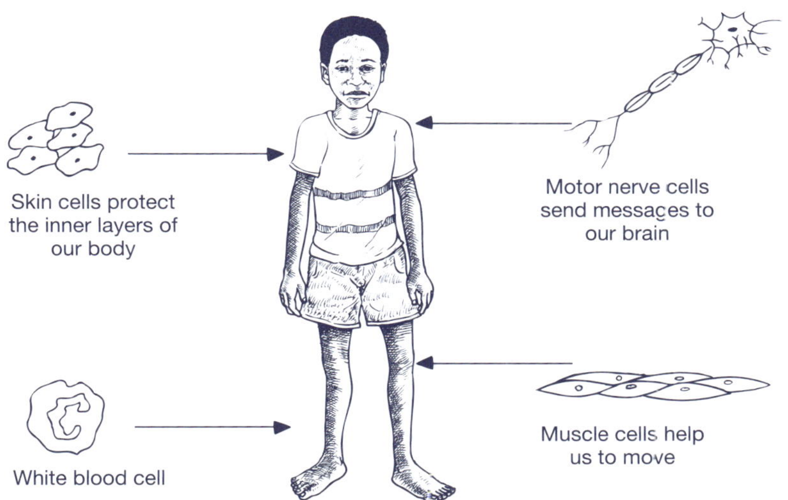

Most plants are also made up of millions of cells.

When we look at cells under the microscope we can see a number of similarities and differences. Most cells have a control centre called the **nucleus** that can be clearly seen. The nucleus is surrounded by a liquid called the **cytoplasm** and a **cell membrane**.

Plant cells and animal cells have several important differences and these are shown in the table below.

The differences between plant and animal cells

Plant cells	Animal cells
Generally shaped like a box	May be any shape and can change shape
Have a hard cell wall made of cellulose	Have no hard cell wall and remain soft
Contain chlorophyll	Do not contain chlorophyll

In many plants the hard cell wall remains after the plant has died and is called **wood**.

This is the reason why wood from trees can be used to make houses, canoes and furniture and why wood can also be burnt as a fuel.

Plants have leaves that are made up of cells.

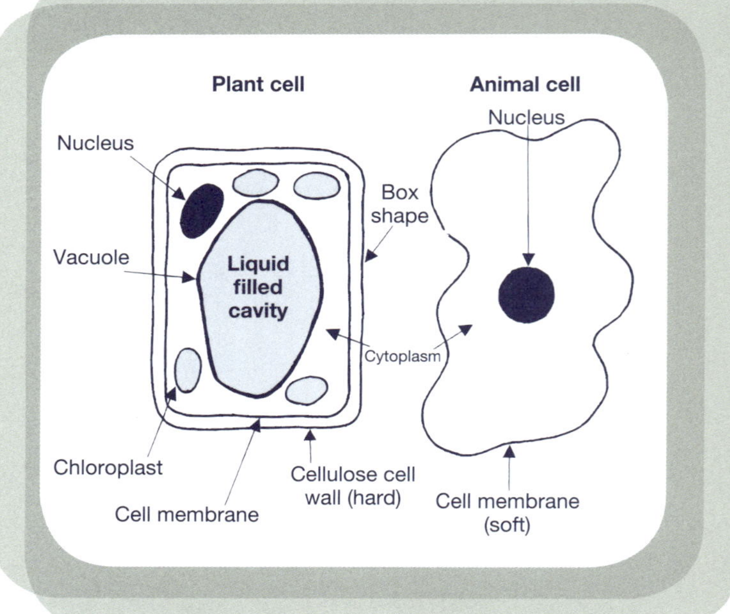

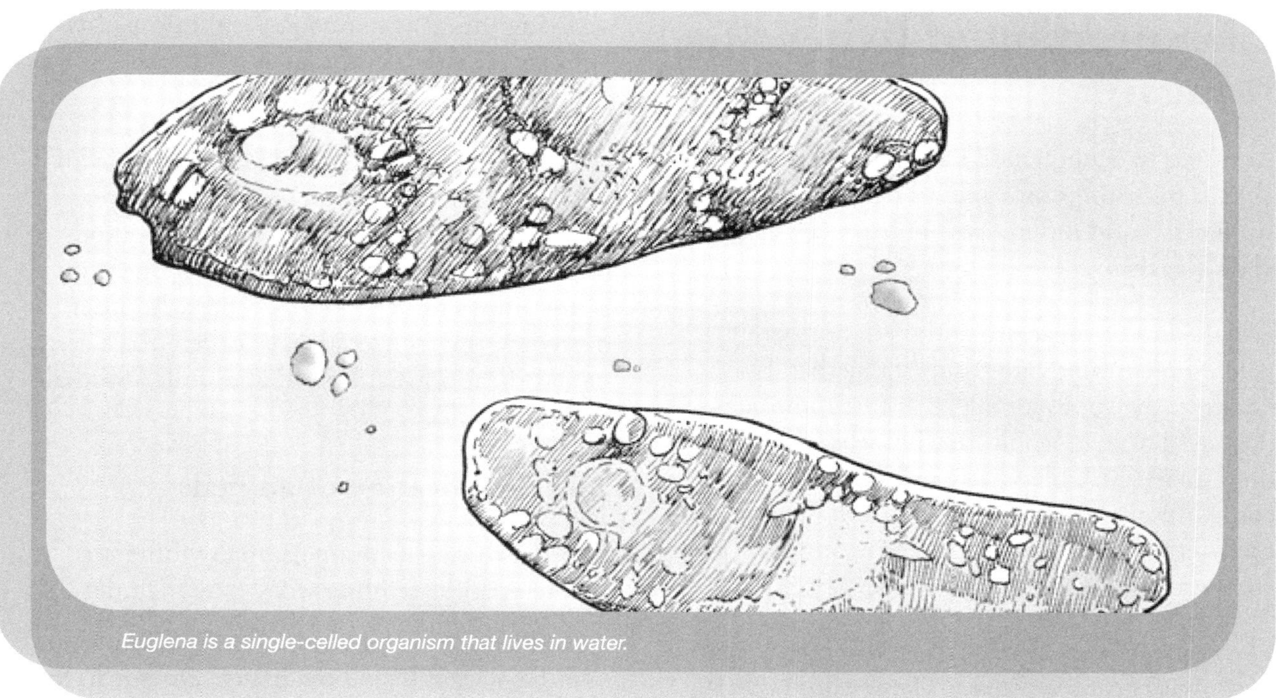

Euglena is a single-celled organism that lives in water.

For you to try

1 **Activity: Making models of cells**

- Collect pieces of cardboard carton, paper, scraps of material, plastic wrapping, plastic containers and anything else that you can find.
- Make a model of a plant cell and an animal cell.
- Make labels for your model to show the cell wall, nucleus, cell membrane and cytoplasm.
- Show and tell others about your models.

2 If your school has a microscope, try the following.

Investigation: Looking at onion skin

- Carefully peel a small piece of very thin skin from the inside layer of an onion.
- Place it on a glass slide and add a drop of water to stop it from drying out.
- Examine the onion skin under low power magnification and try to identify cells.
- Make a drawing of what you can see and label any cell structures.

Classification of living things

At home, at school and at work it is useful to be able to put things into groups depending on their similarities and differences. This process is called **classification**.

Objects can be organised and used more easily when similar things are grouped together. For example, the goods for sale in a store must be organised so that people can find what they want on the shelves. A till or cash register has separate compartments for K20, K10, K5 and K2 notes, each with its own spring clip. There are also separate compartments for each of the types of coins.

Using a system of classification also means that new objects can be fitted in. For example, PNG replaced Australian dollars with kina and toea in 1975. Since then three different notes and one coin have been introduced: the K20 note in 1978, the 50t coin in 1980, the K50 note in 1989 and the K100 note in 2005. These new types of currency can be fitted in with the others. For example, the 50t is a coin so it is grouped with the other coins, and is worth more than 20t but less than K1, so it fits in between these two.

Scientists also look for similarities and differences and put things into different groups according to their **characteristics**. An easy way to classify all living things is to divide them into plants and animals as shown in the table below.

People in different parts of Papua New Guinea have also developed their own way of putting living things into groups. However, different communities use different methods of classification and the language of one group is not always understood by others. In order to overcome this, scientists have developed a system of classification that is used and understood all over the world.

The classification of animals

In order to classify animals, many different characteristics are observed. These include:

- Cell and body structure: How big is it? Does it have legs? If so, how many? Does it have wings, fins or a tail?
- Body covering: Does the animal have scales, feathers or fur?
- Reproduction: How does the animal reproduce? How many young does it produce?
- Movement: How does the animal move? Can it walk, fly or swim?
- Food gathering methods: How does the animal feed itself? What does it eat?
- Method of obtaining oxygen: Does the animal have lungs or gills?

Simple classification of living things

Plants	Animals
May be a single cell or have many cells	May be a single cell or have many cells
Cells with hard walls of cellulose	Cells with no cellulose
Contain chlorophyll (except fungi e.g. moulds, mushrooms)	Contain no chlorophyll
May move a little	Usually free to move about

For you to try

Activity: Animals for classification

1 Look at the animals below.

2 Make a card for each animal like the example below. Each card should list five characteristics of the animal.

> **Pig**
> 1 four legs
> 2 curly tail
> 3 long nose or snout
> 4 tusks coming out of mouth
> 5 dark, hairy skin

3 Look for a characteristic that most of the animals have. Sort your cards into piles according to this characteristic, e.g. number of legs. Make a list of the animals that you have put into this pile.

4 Repeat **step 3** for other characteristics. Make a list of the animals in each pile for each characteristic that you choose.

5 Try putting the animals into groups where they have two characteristics that are the same. Does this make it easier or more difficult to put the animals into groups?

The classification of plants

In order to classify plants we look to see how simple or complex they are. Most of the plants we know are from a group known as **flowering plants** which are very complex. Flowering plants have many different parts and each part has a special job or function. For example, the flowers have the special job of producing the seeds. The seeds will grow into new plants when the conditions are suitable.

There are five other groups of plants. Some of these plants are very simple. Simple plants like algae need to be surrounded by water in order to keep them alive. The more complex plants like conifers and flowering plants do not need a constant source of water. Many flowering plants have special features that help them survive in very dry conditions. For example, they may have many deep roots, and special leaves that reduce the loss of water.

Although the large flowering trees and the conifers can grow very tall, most of the other groups of plants are small and grow to a height of less than one metre. The six groups of plants are shown in the table on the next page.

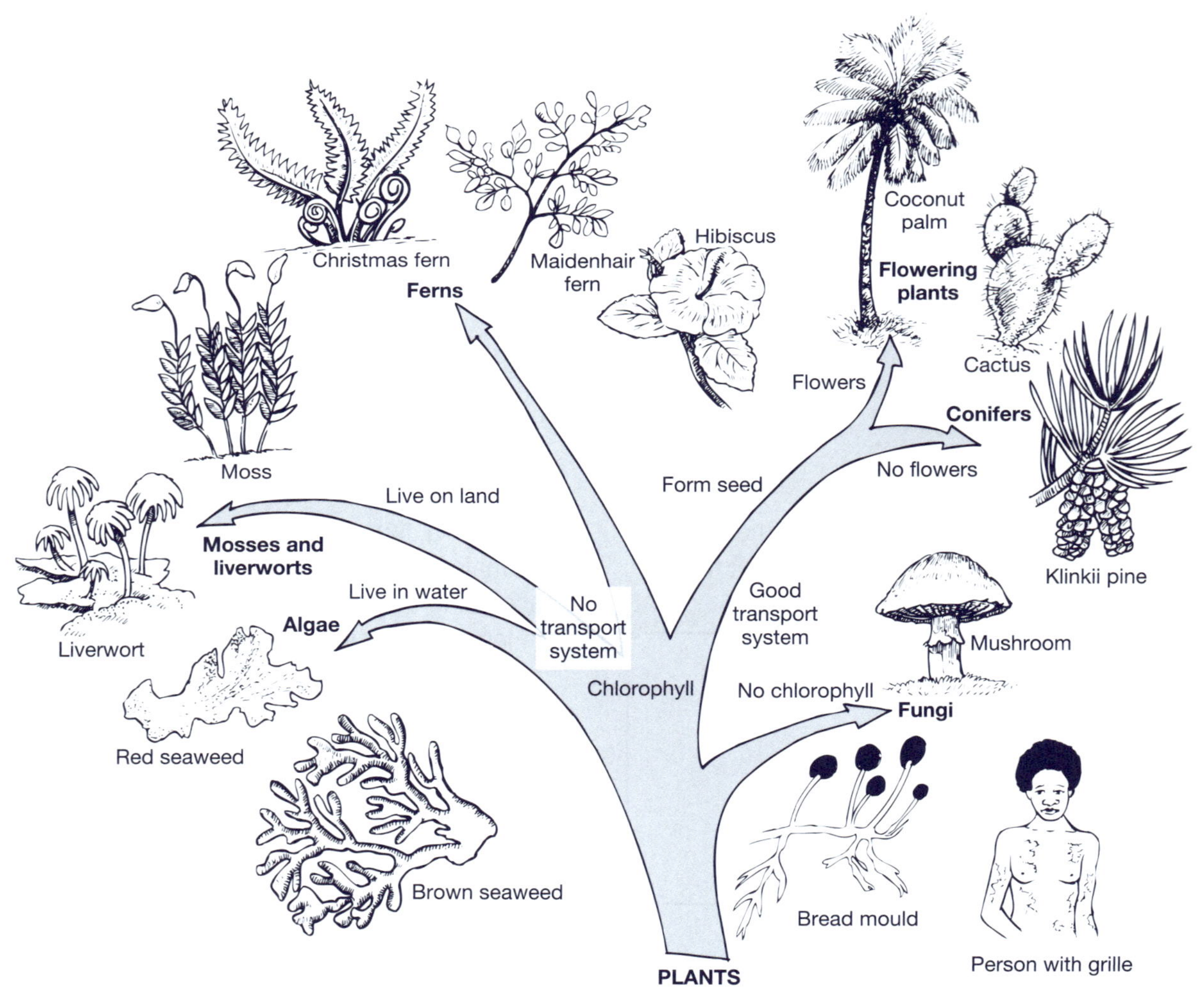

The classification of plants

Main plant groups

Plant group	Examples
Algae	Seaweed
Mosses and liverworts	Funaria moss, liverwort
Ferns	Staghorn fern, bird's nest fern, Salvinia, edible ferns
Conifers	Klinkii pine
Flowering plants	Orchid, hibiscus
Fungi	Mushroom, bread mould, grille

For you to try

1 Investigation: Looking at leaves

a Collect a variety of leaves with different shapes and sizes and a ruler.

b Classify your leaves by size. Measure each leaf and place all leaves larger than 10cm in one pile and all the smaller leaves in another pile.

c Now classify your leaves by shape. Divide the smaller leaves into two piles: those that have a regular shape and those that do not have a regular shape.

d Classify the larger leaves in the same way.

e How many piles of leaves do you have now?

f Think of some other characteristics of your leaves that you could use to classify them differently.

2 Copy and complete the table on the next page using your own knowledge and the information in the diagram on the classification of plants. A

Classifying leaves

Some of the information has been completed for you.

Type of plant	Characteristics			
	Chlorophyll	Transport system	Where does it live?	Flowers
Flowering plants	Have chlorophyll	Have good transport system		Have flowers
Conifers				
Fungi				
Ferns		Have good transport system		
Mosses and liverworts	Have chlorophyll			
Algae			Live in water	

3 Make list of ten plants and say how they are useful to the community.

Levels of classification

The features or characteristics of living things can be used to further classify them into separate groups. In a classification system each group must be different to all other groups and it can then be given its own name. In this way the members of each group will have similarities with other members of the group, but will have differences with members of other groups. For example, plants and animals can be divided up into many more groups. Plants can be divided into **flowering** and **non-flowering** plants and each of these two main groups can be divided into smaller groups, which is the next level of classification. Animals can be divided into those without backbones, which are called **invertebrates**, and those with backbones, which are called **vertebrates**. Each of these two main groups can again be divided into smaller groups. The first levels of classification of living things are shown in the table on page 35.

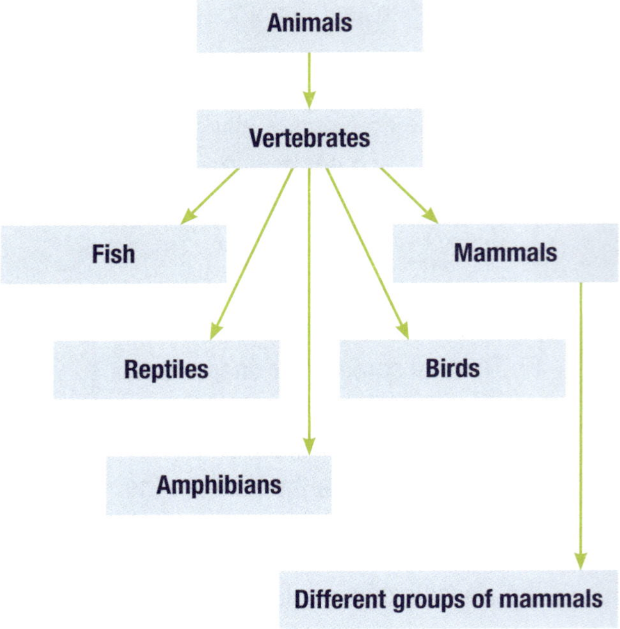

The classification groups for vertebrates

Living things			
Plants		**Animals**	
Flowering plants	**Non-flowering plants**	**Animals without backbones**	**Animals with backbones**
	Algae	Sponges	Fish, e.g shark
	Mosses and liverworts	Jellyfish	Amphibians, e.g. frog
	Ferns	Worms: flatworms, roundworms, segmented worms	Reptiles, e.g. crocodile, turtle
	Conifers	Sea urchin, sea star	Birds, e.g. cassowary,
	Fungi	Shellfish	Mammals, e.g. human, wallaby
		Spiders, crabs and insects	

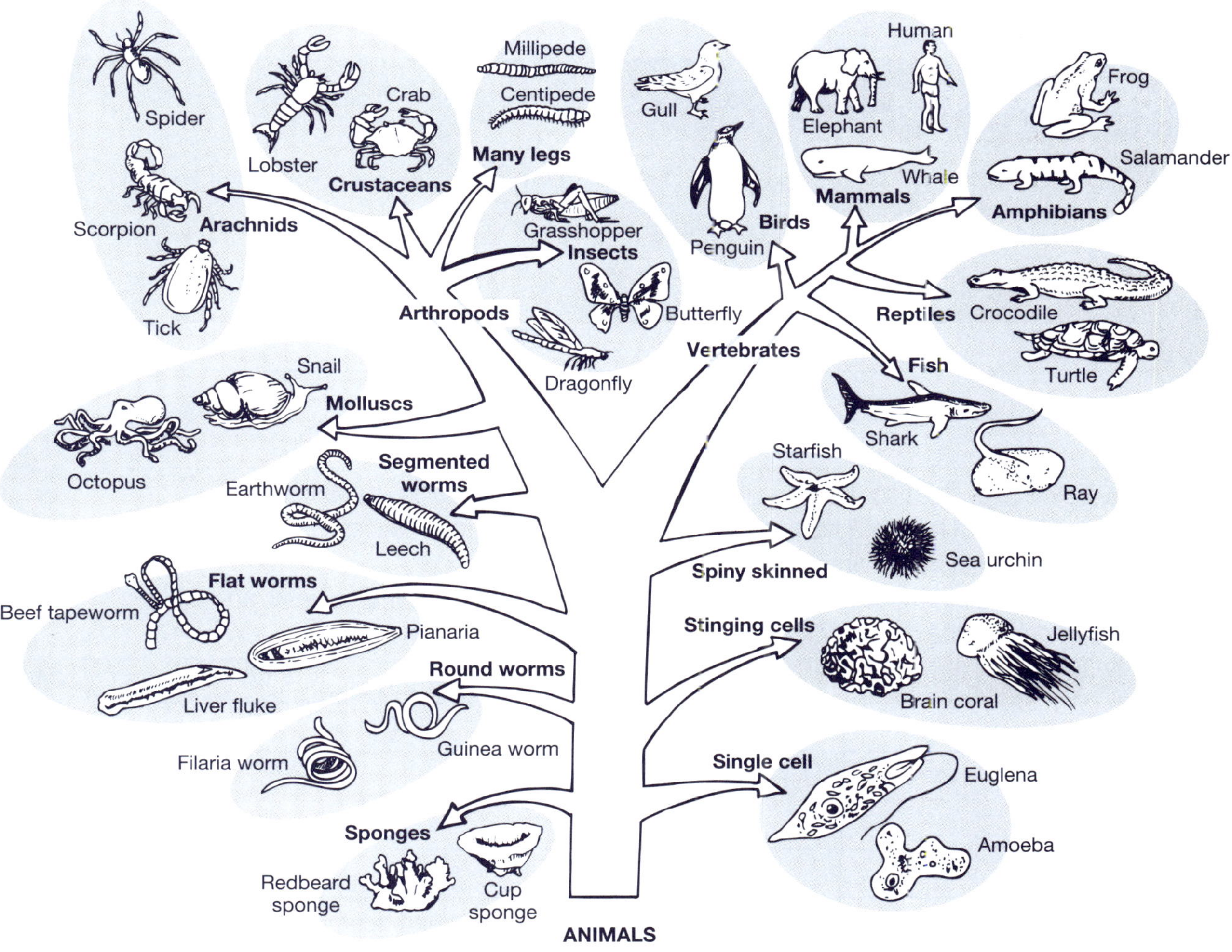

The main animal groups

Using a key

Sometimes we use a **key** to help us to sort things into groups or to classify them. A key is a series of pairs of questions that we ask about the thing that we are trying to classify. The questions must be 'either–or' questions, so that the answer is clearly 'yes' or 'no' to each question or choice. Each answer will lead to another pair of questions until we get to the final answer. For example, the key on this page can be used to classify vertebrates into five groups.

To use the key, start at the top and read across the table from left to right. Start with **1** and choose **a** or **b**. After answering the first pair of questions then answer the second pair of questions, and so on. Continue until you have sorted the unknown animal into one of the five groups.

By following the four steps in this key it is possible to classify any unknown vertebrate and work out whether it is a fish, amphibian, reptile, bird or mammal.

Key for the classification of animals with backbones

1	a b	feathers present? → no feathers →	birds go to 2 →
2	a b	hair or fur present? → no hair or fur present →	mammals go to 3 →
3	a b	fins present? → no fins →	fish go to 4 →
4	a b	dry skin → moist skin →	reptiles amphibians

For you to try

1 Copy and complete the following table of living and non-living things that are found around your home or school.

Living and non-living things	
Living things	**Non-living things**

2 How do we know that something is living? What characteristics do you look for?
3 Why do think that it is an advantage for the mimosa to fold up its leaves when it is touched? How does this help the plant?
4 Copy and complete the following table to show the different ways that animals are able to move; try to include many different types of movement. A

Animal	Ways of making themselves move

5 Use the key to classify some animals with backbones that you find near your home or school.

Cells, tissues and organs

Plants and animals are made up of millions of **cells**. Because we are also animals our bodies are also made up of millions of cells. Cells that are similar to each other are found in groups called **tissues**. The cells that are found in a tissue have a similar appearance and have the same job or function. Some examples are shown below.

Cells	Tissue	Job or function
Skin cells	Skin tissue	Stop bacteria entering the body
Muscle cells	Muscle tissue	Allow the body to move or pump blood
Bone cells	Bone tissue	Support the body; make blood cells

An **organ** is a part of the body that is made up of two or more different kinds of tissue that work together to carry out a particular job. For example, the heart is an organ that is made of muscle tissue that can pump blood. The lungs contain lung tissue that can swap oxygen and carbon dioxide between the air and the blood.

A **system** is made up of a group of organs and tissues that work together to carry out a particular job. For example:

- The **circulatory system** is made up of the heart and blood vessels and carries blood to all parts of the body.
- The **nervous system** is made up of the brain, spinal cord and nerves which carry messages to all parts of the body.
- The **respiratory system** is made up of the lungs, breathing tubes and some muscles and allows gases to enter and leave the body.

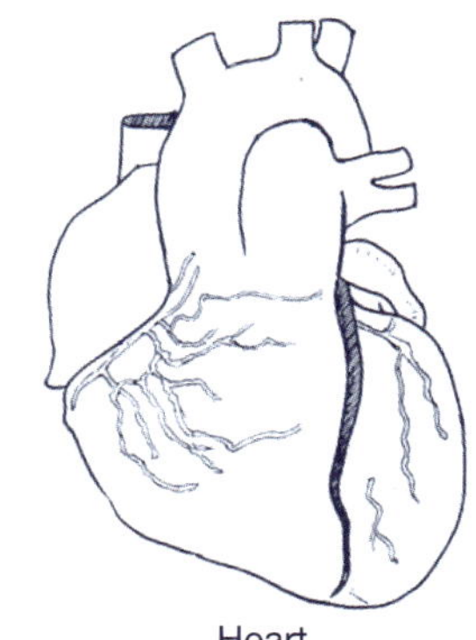
Heart

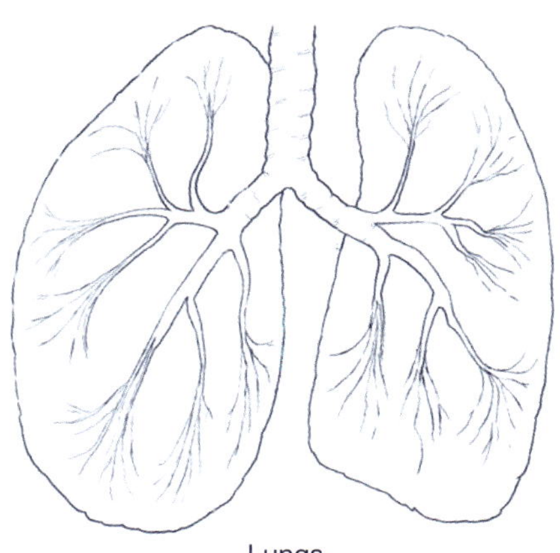
Lungs

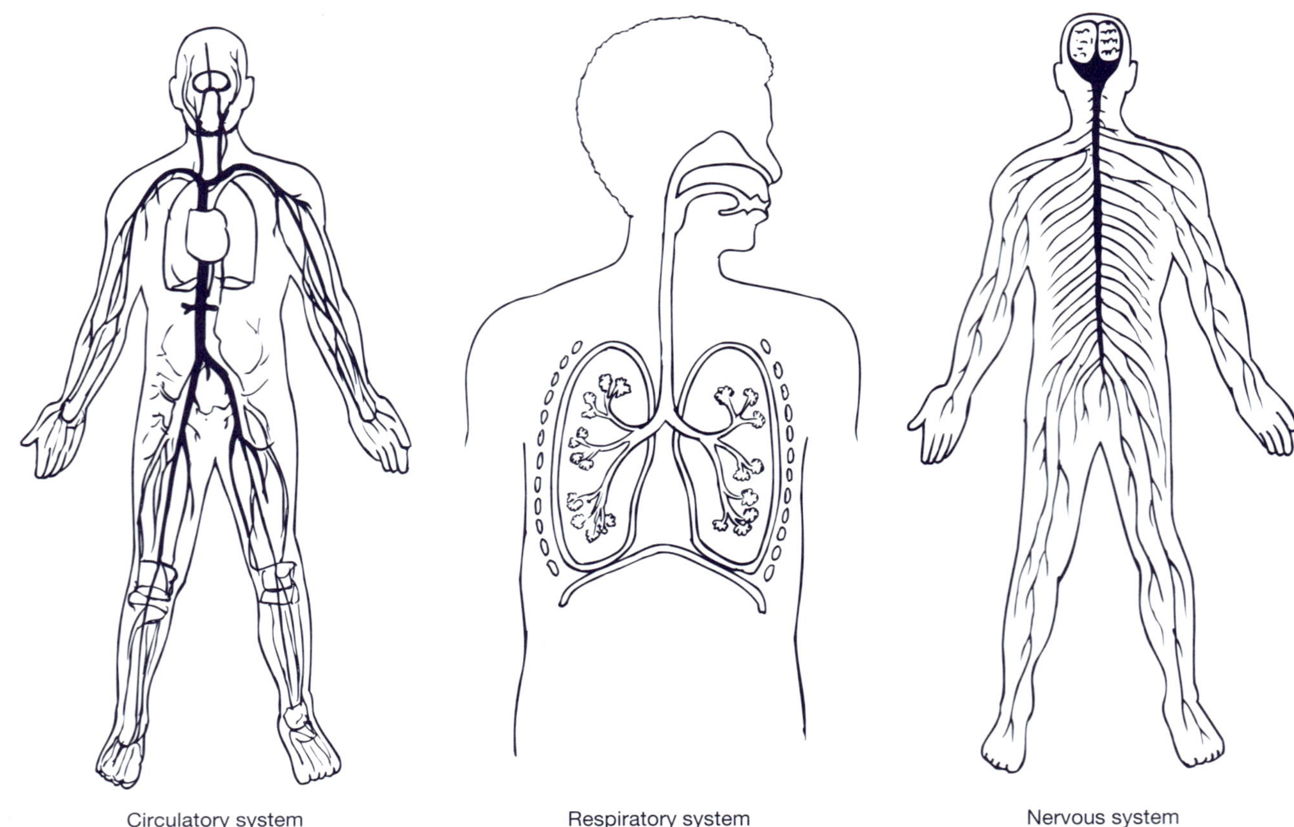
Circulatory system

Respiratory system

Nervous system

Living things are also called **organisms**. All the systems work together in an organism to help the organism to survive in the environment. For example, an animal needs both a skeleton and a muscle system in order to be able to move. The muscles must pull on bones in order to make the body move. As well as allowing the body to move, the skeleton also gives support and protection.

Sense organs and senses

In animals, some organs help to provide information that the organism needs in order to be able to live and survive in the environment. For example, humans use their senses to find out what is going on around them.

- We use our **eyes** to see the shape and colour of things and to read about interesting and important things.
- We use our **ears** to listen to music, to hear people speak and to know if some danger is coming towards us.
- Our sense of **taste** tells us if food is too salty or too sweet.
- Our sense of **smell** can tell us if food has started to go rotten and also helps us enjoy the flavour of food.
- Our sense of **touch** can tell us whether things are rough or smooth, hot or cold.
- We can also feel the heaviness of things and judge the size of things by touching them with our hands.

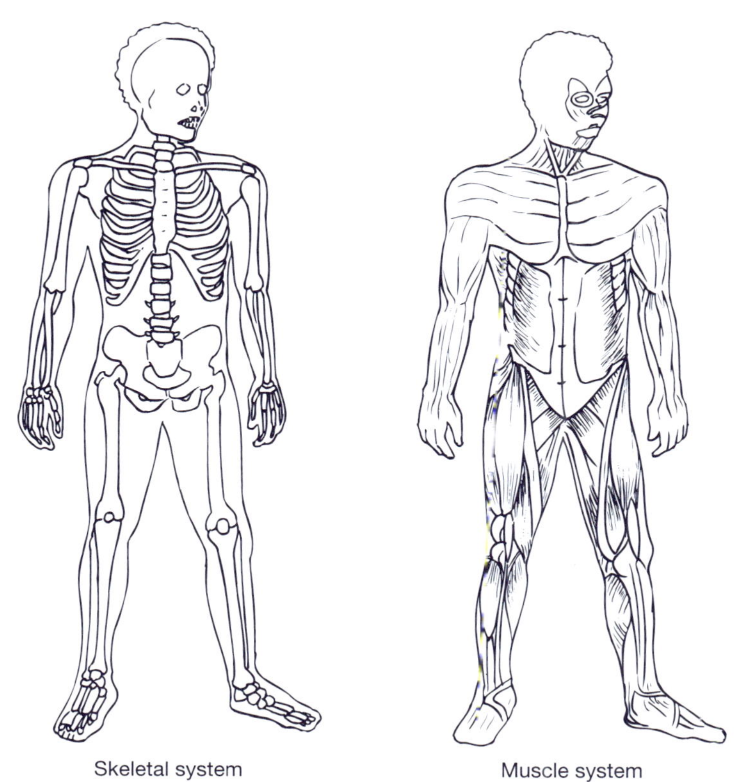

Skeletal system   Muscle system

Touch

Sight

Smell

Taste

Hearing

BARRAMUNDI

BARRAMUNDI

Using our five senses

The sense of sight

Sight is the ability to see things. We use our eyes to detect the light coming from an object. With our eyes we can see large things that are a long way away and very small things that are close by. We can see the movement of other people, what they are doing and what they are wearing.

Sight is also important in other animals. For example:

- Many birds have good eyesight for hunting. For example, birds of prey like eagles have eyes that point forwards which are good for judging the distance of objects. Other birds and mice and rats have eyes on the side of the head which give good all round vision.
- Flies have special eyes that are very good at detecting movement, which is why they are so difficult to catch.
- Fish have eyes that allow them to see under water.

Birds of prey like eagles and owls have eyes that point forwards.

Insects have special eyes made of many small parts.

Fish have eyes that allow them to see under water.

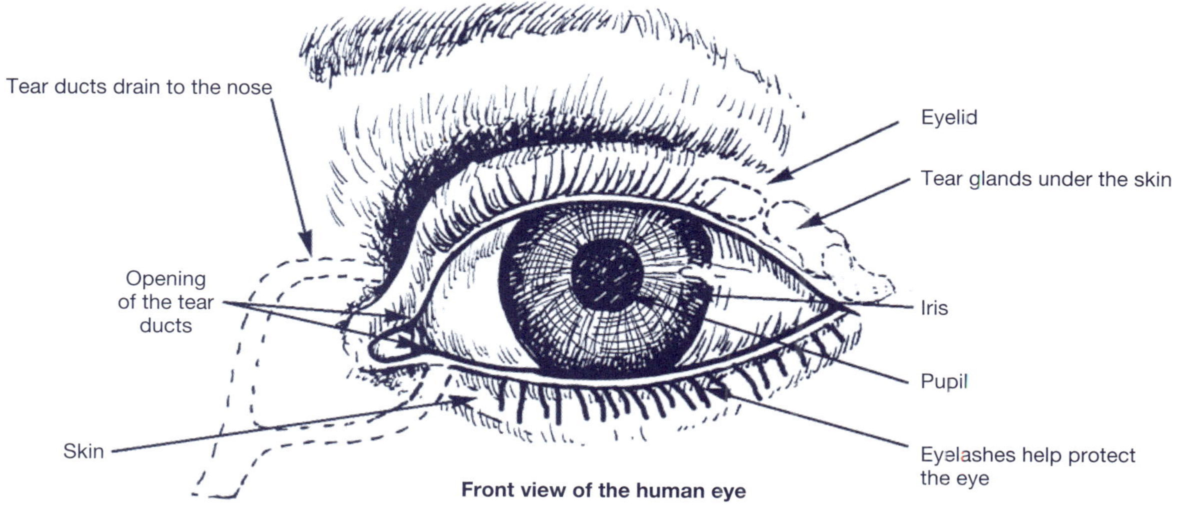

Front view of the human eye

For you to try

Investigation: Are two eyes better than one?

SR

1. Collect two pencils.
2. Work with a partner. One person does the experiment first while the other observes, then swap places.
3. Hold one pencil in each hand at arm's length with the points facing each other.
4. Close your left eye and slowly bring the pencils together and try to make the points touch.
5. Describe what happens.
6. Repeat with your right eye closed.
7. Repeat with both eyes open.
8. Swap over so that the other partner can have a go.
9. Do both partners get the same result?
10. Write a report of what happened and include a sentence about the difference between using one eye and two eyes. Are two eyes better than one? Why?

The sense of hearing

Hearing is the ability to detect sounds. We use our ears to hear the sounds produced around us. With our ears we can detect noises and music. We can gather information from our surroundings and by listening to what people are saying.

Other mammals, birds, reptiles and amphibians also have ears. These animals can also make sounds and so use the sense of hearing as a way of communicating.

Dogs have a better sense of hearing than humans.

- Dogs and cats can hear sounds that humans cannot hear. They can move their ears to work out the direction from which the sound is coming. Humans cannot move their ears in this way but instead turn the head to face the direction of the sound.
- Birds can find a mate and tell other birds to keep away using their songs.
- Frogs and insects can find each other by using their sense of hearing to listen to the sound they make.
- Whales make noise under water and can hear each other over very long distances.

For you to try

1 Investigation: Sound in a tube

a Collect a long tube made from cardboard or a rolled-up newspaper. You will also need a watch or small clock that ticks.

b Hold the watch near your ear and gradually move it away until you cannot hear it any more. How far do you have to move it?

c Hold one end of the tube against your ear and hold the watch in the opening of the other end of the tube.

d Describe and explain what happens.

2 Investigation: Where's the sound?

a Collect a jar containing dried seeds or small stones, a piece of material to make a blindfold, and some sheets of paper.

b Work in groups of three and take turns at making the sound, listening and recording the results on the sheet of paper.

c The first person sits in a chair and puts on the blindfold. Cover your left ear with your hand.

d The second person should stand about one metre away and rattle the jar. The first person then points to where they think the sound is coming from.

e The third person records on a circle drawn on a sheet of paper the position where the sound was made and whether the person could hear the sound.

f The second person will now move to different positions in a circle around the first person. The jar is rattled in each of these positions and the result is recorded.

g Now cover your right ear and repeat the same steps.

h Repeat the experiment holding a paper cone to each ear instead of covering your ears. What difference does this make?

The sense of smell

Smell is the ability to detect odours. Our noses contain special sensors that can pick up the smell of small amounts of gases that we breathe in. Our sense of smell can be used to detect danger, such as fires or leaking gas. We can also smell if something is going rotten or decaying. For example, we can tell if food is going bad or if an animal has died.

Many animals have a sense of smell. For example, cats and dogs have a well-developed sense of smell and can detect odours that humans cannot smell. Sharks can smell a very small amount of blood in the water and insects like butterflies can find a mate using smell.

Sharks and butterflies have a very good sense of smell.

For you to try

Investigation: How does smell affect taste?

SR

1 In this investigation you will stop your sense of smell so that you can see how it affects the way you taste things.
2 Collect a piece of three different kinds of raw food such as banana, pawpaw, watermelon etc., three bowls and spoons, and a piece of material for a blindfold.
3 Grate or chop each of the three foods to make very small pieces and place in a bowl.
4 Blindfold yourself and hold your nose tightly.
5 Ask another student to feed you a spoonful of each food.
6 Can you recognise which food you are tasting?
7 Swap over to let the other person have a go.
8 Discuss your results with other groups.
9 Write up your investigation to say what you

The sense of taste

Taste is the ability to recognise different kinds of food in the mouth. Our tongue has taste buds that can detect only four different tastes: sweet, sour, salty and bitter.

Food has many different **flavours** because when we eat food the taste buds in the tongue can taste the food and at the same time the nose can smell the food in the mouth. When you have a cold and your nose is blocked, you can still taste the food but it does not seem to have much flavour and is less enjoyable. When this happens we realise that the flavour and our satisfaction or enjoyment depend on the smell, not just the taste of the food.

Our sense of taste tells us if something is too sweet, sour, salty or bitter and so is not good to eat. We can also tell if something has gone bad by the taste.

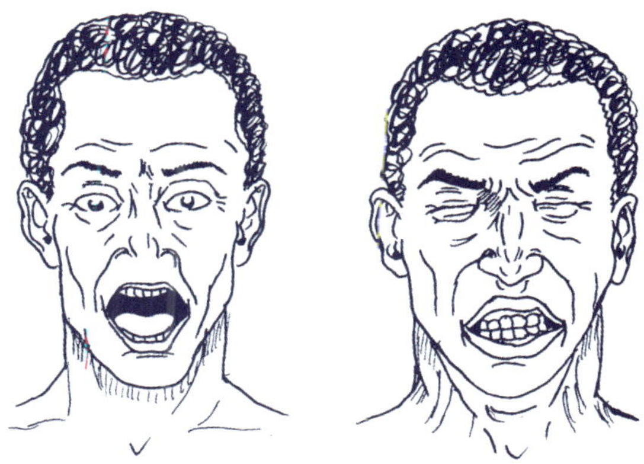

People sometimes pull a face when they taste something too sweet, too sour, too salty or too bitter.

Other animals also have a sense of taste and may prefer some flavours more than others. Flies use their feet to taste their food.

For you to try

Investigation: Making a tongue map

SR

1. Collect four small cups, some sugar, a lemon or some vinegar, some salt, one chloroquine tablet and four small plastic drinking straws. You can reuse the small bendy straws from a fruit drink box if you wash them carefully first.
2. Label small four cups or similar containers **i**, **ii**, **iii** and **iv**. Add the following to each cup as follows:
   - **i** sugary water (sweet)
   - **ii** lemon juice or vinegar (sour)
   - **iii** salty water (salt)
   - **iv** one crushed chloroquine tablet in a small amount of water (bitter)
3. Using a drinking straw (or a clean strip of paper) place a few drops of liquid **i** on your tongue. Try different parts of the tongue, like the tip, the back and the sides.
4. Draw a map of the tongue and mark on the map where you taste the sugar water most strongly.
5. Rinse your mouth with water and repeat with the other three liquids.
6. Write up your investigation to show what you did and what you found out. Include your map of the tongue to show which parts of the tongue are most sensitive to different tastes.

7 Note that some tastes may be detected in more than one place and the tip of the tongue detects two tastes.

8 Share your results with other groups. What differences did you notice?

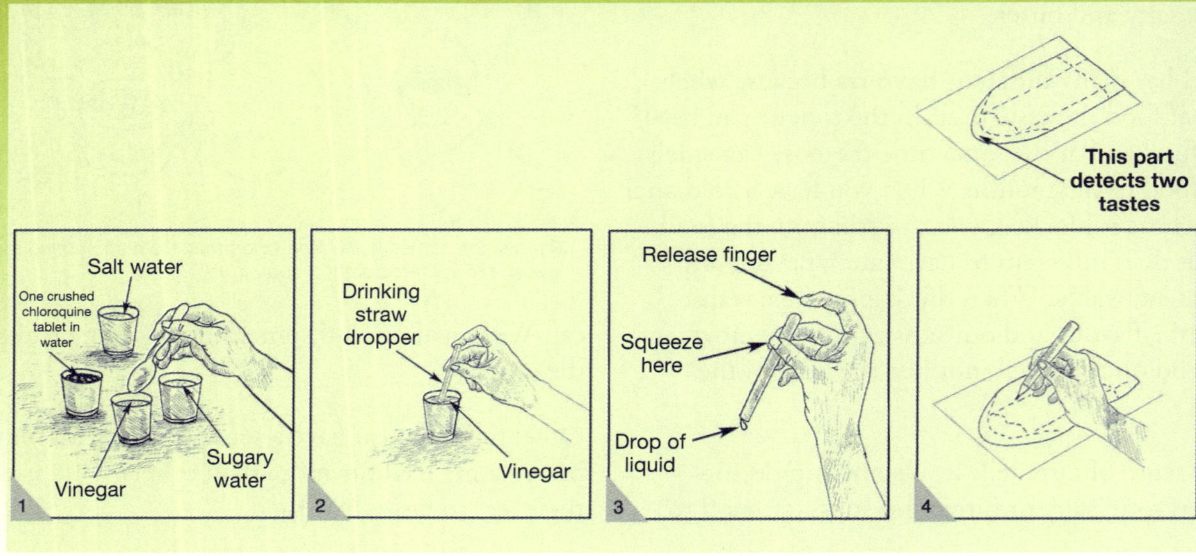

The sense of touch

Our sense of touch can be used to detect the hotness or coldness of things. For example, our hands can easily tell if a cooking pot is too hot to carry or not. Our sense of touch also helps to detect the mass or heaviness of different objects. For example, when our clothes get wet in the rain, they feel heavier than when they are dry.

Reading Braille

Humans have such a well-developed sense of touch that they can understand what something is like by lightly touching it. This sense becomes highly developed in people who cannot see because they can learn to read by using their fingertips to feel specially patterned bumps pushed into paper. The Mount Sion Institute in Goroka can prepare textbooks and examinations in Braille for students who cannot see.

Fish have an organ inside the body which allows them to detect if something is moving towards them. This is the reason why it is hard to catch a fish with your hands.

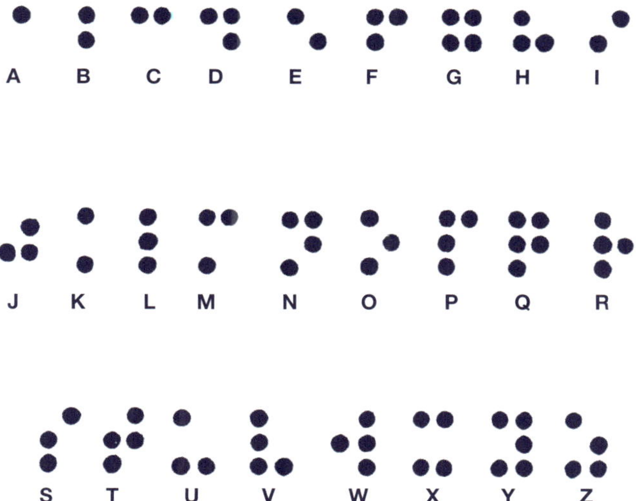

The Braille alphabet

For you to try

1 Investigation: Getting the point

SR

- **a** Collect two sharp pencils, a rubber band and a piece of material to make a blindfold.
- **b** Join the two pencils together with the rubber band so that both points are level with each other
- **c** Blindfold yourself. Then ask your partner to touch very lightly in the places shown in the diagram.
- **d** Each time you must say whether you can feel one point or two points.
- **e** Swap over and compare your results.
- **f** Discuss your results with other groups.
- **g** Write a report of your investigation to say what you did and what you found out.

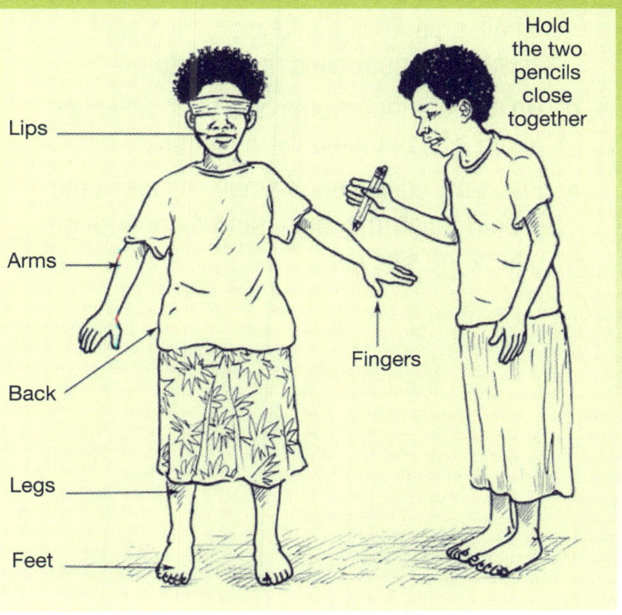

2 Investigation: The touch test

Work in groups for this investigation. SR

A T-shirt touch-bag

- **a** Make a 'touch-bag' by folding over the hem of an old T-shirt and sewing it. Sew the neck together but leave the sleeves open. If you want to wear the T-shirt again later you can use big stitches so that you can easily undo them.
- **b** Another way to make a 'touch-bag' is to use an old flour sack or rice bag.
- **c** Collect a number of small objects that can easily be touched and held in the hands. For example, various coins, a two Kina note, a piece of paper or plastic exactly the same size as a two Kina note, a marble, a stone, a piece of chalk, a big seed, a flower, a small sweet potato, a nail, a paperclip, a key, a pencil, an eraser, a rubber band, a lolly, etc.
- **d** Do not let other groups see your collection of objects, but keep them hidden.
- **e** Place the collection of small objects in the T-shirt touch-bag by putting them through one of the sleeves.
- **f** Invite your classmates from another group to come and put their hands in the touch-bag and find out what objects are inside.
- **g** How many objects did they guess correctly? Which objects are most difficult to guess? Could they tell the difference between the two Kina note and the piece of paper or plastic that was the same size?
- **h** Write a report of your investigation to say what you did and what you found out.

3 Investigation: Earthworm responses

Work in groups for this investigation.

a Collect an earthworm, a shallow dish or bowl, some newspaper and a lemon. Remember that the earthworm is a living organism, so be careful and gentle. Keep the earthworm in a jar of loose damp soil or damp grass until you are ready.

b Place the earthworm on damp newspaper in the bottom of the dish or bowl.

c Carefully investigate the earthworm's response to the following:

- light: make a 'roof' of newspaper so that the worm can choose to stay in the light or go under the 'roof' where it is darker
- touch: use the tip of a pencil
- substances: use a piece of paper or cloth soaked in lemon juice.

d Test each end of the earthworm and the sides (for touch and substances).

e Describe how the earthworm responds each time.

f Are some parts of the earthworm more responsive than others?

g When you have finished, put the earthworm back in the soil.

Finding out about earthworm responses

4 Discuss in small groups why it is important for living things to have sense organs. What would happen if they did not have sense organs?

5 Copy and complete the following table to show the way in which the senses are used to help each animal live in its environment. A

How senses help animals to live in the environment

Sense	Humans	Choose another animal:
Sight		
Hearing		
Taste		
Smell		
Touch		

Show your table to others in your class. Notice any similarities and differences.

Ecology, relationships and interactions

Living things are usually well suited to the place in which they live. In other words, living things are **adapted** to their way of life. When we look at living things we can see that they interact with each other and with their non-living surroundings. The study of the relationships and interactions between living things and their surroundings is called **ecology**.

Every organism has to satisfy its needs if it is to survive and reproduce. For example, all living things must have food, water and shelter. Animals must also try to make sure that they are not eaten by other animals. Food is probably the most important need and feeding is the most important interaction between living things.

Plants are able to make their own food and so they are also known as **producers**. Plants can produce food from carbon dioxide, water and minerals in a process called **photosynthesis**. The energy they receive from the sun is trapped in the food they produce. Plants use the food they produce to grow and reproduce.

Animals cannot make their own food and so must find and then eat food. Because animals eat or consume their food they are called **consumers**. Many animals obtain their food by eating plants. Animals that eat only plants are called **herbivores**. Possums, tree kangaroos, grasshoppers and butterflies are herbivores.

Some animals do not eat plants but they eat other animals and are called **carnivores**. Snakes, hawks, sharks and spiders are examples of carnivores.

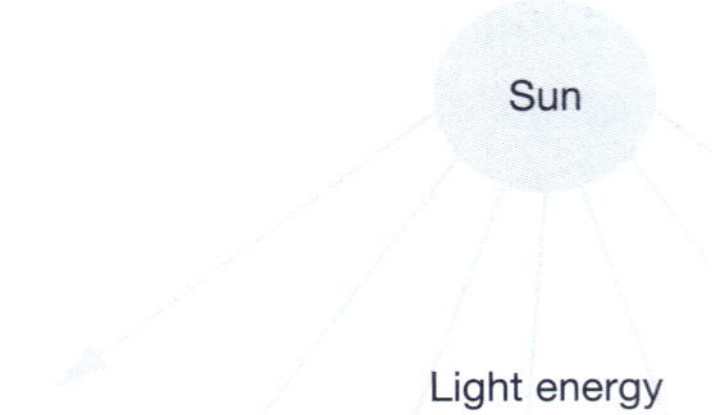

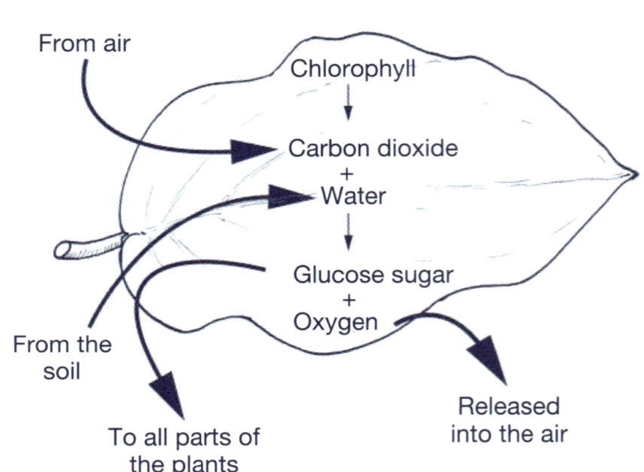

The process of photosynthesis

A few animals eat both plants and animals and are called **omnivores**. For example, honeyeater birds eat insects when they cannot find enough flowers that provide them with nectar. Humans and pigs are also omnivores.

Some animals also feed on dead organisms and are called **scavengers**. Crows, some crabs, some rats and some ants feed like this. Scavengers are useful because they help to clean up dead animals.

The crocodile is a carnivore and the cow is a herbivore.

The pig is an omnivore.

Crows are scavengers and help to clean up dead animals.

As well as the feeding relationship, living things interact in providing shelter for other living things. For example, birds and cuscus shelter in trees.

Living things also depend on their non-living surroundings. Plants need water, energy from sunlight, carbon dioxide from the air, nutrients from the soil and shelter from their surroundings. Animals also need water, sunlight, oxygen from the air and shelter from their surroundings.

Food chains and food webs

When we look at the feeding relationships between living things in an area we can see that there is a chain of organisms in which one feeds on the other. For example, grass is eaten by a grasshopper, which is then eaten by a frog, which might then be eaten by a snake. This feeding relationship is called a **food chain**. We can see from this relationship that all the organisms in a food chain depend on the producer, which is a plant. Plants are the most important organisms because they are the only ones that can produce food. Plants change the energy from the sun into food which is another type of energy. Animals make use of this energy when they eat plants and when they eat other animals.

Each organism is usually in more than one food chain. For example, the frog may be in another food chain:

pond weed → water beetle → frog → cattle egret

Snake (Third-order consumer)
Frog (Second-order consumer)
Grasshopper (First-order consumer)
Grass (Producer)

A food chain showing the way that energy is passed along.

The arrows show the direction in which the food energy is moving along the chain. There are usually many food chains in an area and they are linked together to form a **food web**. Food webs are usually very complicated.

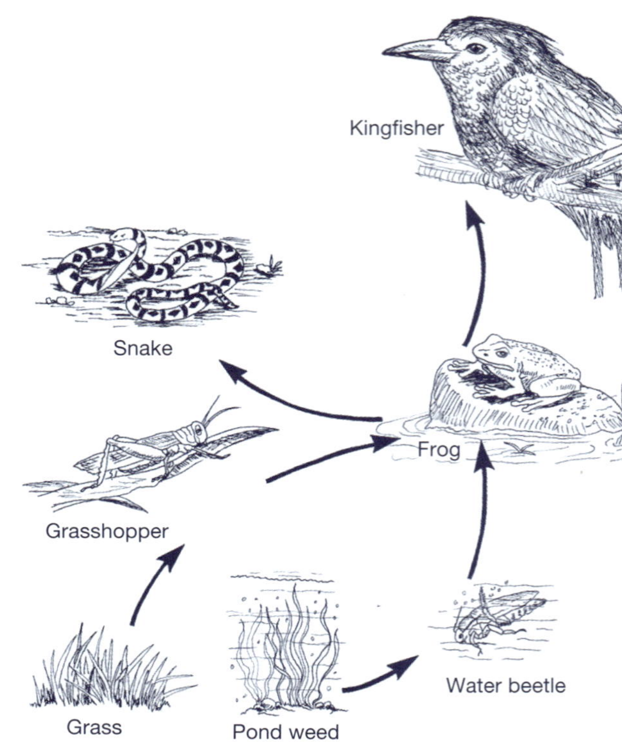

Two food chains with a frog in both, forming a simple food web.

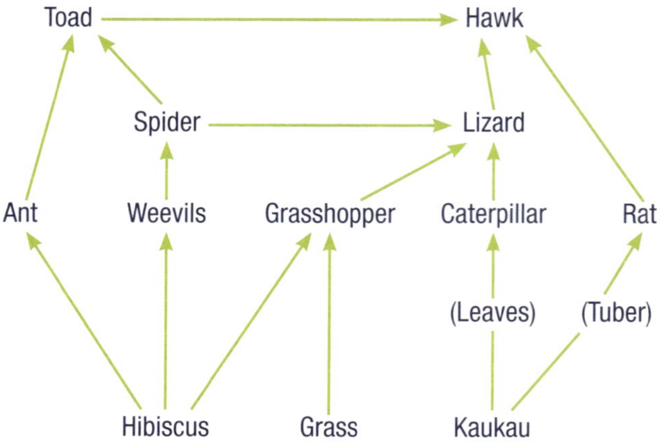

Part of a food web showing how some organisms are in many food chains.

For you to try

1. Explain what is meant by a food chain. Give an example.
2. Explain what is meant by a food web.
3. Use the diagram of the food web shown above to answer the following.
   - a Name a herbivore.
   - b Name a carnivore.
   - c Write down three different food chains containing the hawk.
   - d List three food chains, one containing three organisms, one containing four organisms and one containing five organisms.
   - e In how many food chains is the hawk?
4. Why are plants known as the most important food source in the environment? How do they produce their food?
5. **Investigation: Looking at feeding relationships**
   - a Go outside and observe the different living things that you can see in your area. If you live near the sea, a lake or a river you can choose one of these places, but you must take care near water.
   - b Make a list of the plants and animals that are living together and how each obtains its food.

     You could use a table like the one below.
   - c Draw a number of food chains to include the plants and animals that you have found.
   - d Draw a food web that includes the food chains that you have found in the area.
6. Make a list of all the food that you have eaten in the last two days.
   - a Divide the list into plant foods and animal foods.
   - b For each of the animal foods that you have eaten, make a list of the foods that the animal might have eaten. For example, a chicken might have eaten insects, seeds and leaves.
   - c Write down three different food chains that begin with a plant and end with you.

Name of organism	What does it feed on?	What feeds on it?

Traditional gardening or subsistence agriculture

PD SS

Many of the relationships and interactions in ecology can be understood by looking at the example of traditional gardening or subsistence agriculture.

Crops that are planted in gardens need good soil containing nutrients, sunlight and water. People know this and so garden sites are chosen carefully. When people make gardens they often go through the following steps.

- Most of the trees and other plants which are growing in the area are cut down.
- The animals that depend on these plants for food and shelter have to move to another area or they may die.
- After food crops are planted the bush starts to grow again because these plants are better suited to the local conditions than the crops.
- People help the crops to grow better by regularly weeding the garden. In this way people change the relationship between the living things and their surroundings in favour of the crops that they have planted.
- After a few years the crops no longer grow as well as they did previously because nutrients in the soil have been used up.
- People stop using the garden, no more crops are planted and the bush grows back. This is called **bush fallow**. Fallow means that the land is not being used.
- Animals can then return to the area and relationships may again be similar to the way they were in the beginning.
- Over many years, the fertility of the soil slowly improves and the bush may be able to be cut down again and another garden planted.

When the bush is cut down too often in the same area, the soil will gradually lose its nutrients until the bush may not grow back again. The area may then turn into grassland and the soil may never

A garden site is prepared by cutting back the bush.

improve again. If this occurs the original plants and animals that lived in the bush cannot return. However, different plants and animals may move into the area.

This method of gardening shows that people can cause major changes to their surroundings and to the living things in their surroundings. This in turn will change the lifestyle of the people:

- there will be fewer animals to hunt
- their gardens will not produce as much food
- there will be a shortage of timber.

When people stop using a garden the bush is allowed to grow back.

For you to try

1 Find out about subsistence gardening in your area. You could talk to old people or invite one of them to come and give a talk.
   - Do they usually cut down the big bush to make a garden?
   - How long do people wait before they use the same piece of land again?
   - Is the length of the bush fallow getting longer, shorter or staying the same?

2 Make a display poster to show the steps in making a garden or the way that gardens are made in your area.

Summary questions

1 Which of the following best describes the characteristics that we use to tell the difference between plants and animals?

i their movement

ii whether they make their own food or not

iii the structure of the cells

A i only

B ii only

C i and ii only

D i, ii and iii

2 Which of the following best describes the different parts that make an organism, from the biggest to the smallest?

A organs, systems, tissues, cells

B systems, organs, cells, tissues

C systems, tissues, cells, organs

D systems, organs, tissues, cells

3 Which of the following senses are needed to allow us to fully enjoy the different flavours of food?

A taste only

B taste and smell

C taste and touch

D taste and sight

4 Which of the following best describes the job of sense organs?

A provide information about the shape and colour of things

B provide information that helps organisms find food

C provide information that helps the organism live and survive

D provide information so that the organism can avoid danger

5 Which of the following best describes the way that energy moves along a food chain from one type of organism to another?

A producer → scavenger → carnivore → herbivore

B producer → herbivore → carnivore → scavenger

C producer → scavenger → herbivore → omnivore

D photosynthesis → herbivore → carnivore → scavenger

6 A particular food chain contains only three organisms. Which of the following is most likely to show their feeding relationship?

A producer → omnivore → scavenger

B producer → carnivore → scavenger

C producer → scavenger → carnivore

D producer → scavenger → herbivore

7 The passage below is a summary of the main ideas of this chapter. Copy and complete the passage in your book. Using the words in the list, find the words that are missing. You can use each word only once. (L)

animals, backbones, cells, classification, differences, flowering, live, organs, plants, relationship, sense, system, tissues, types

All living things can be placed into groups depending on various characteristics. This is called grouping or ______________. Similarities and ______________ between people and things are used to put them in the same group or in different groups. The two main groups of living things are _____________ and plants. Animals can be divided into two groups: animals without ______________ and animals with backbones. Plants can be divided into two groups called _____________ plants and non-flowering plants. All living things are made of building blocks called _____. Plants and animals have different _______ of cells. Cells are grouped together to form __________ . Two or more tissues together form ________. Tissues and organs together form a _______ that carries out a particular job. Many animals have ________ organs which allow them to find out about their surroundings and to live successfully. Plants and animals ________ and interact together in the surroundings. One important ______________ between plants and animals is the food chain. All animals depend on ___________ for food.

For you to try

Project: studying an old garden

1. Choose an old garden site (or an area which has been cleared of plants but is growing back).
2. Using sticks and string mark out an area of one square metre. This is called making a **sample plot**. Rather than looking at everything in an area we look at a small part or sample and when we understand this part it will help us to understand the bigger area.
3. If possible, repeat this in three other places in different parts of the old garden. Depending on the size and shape of the old garden, one way to do this is to spread the sample plots out in a line with space in between each plot.
4. For each plot, observe and record in a table the plants and animals that are found in the square. Remember some of the animals may be hiding so you may have to look underneath a stone or a log. Try to use the name of each plant and animal, but if not you can just describe its appearance. For example, 'hairy brown caterpillar'.
5. Estimate the numbers or the amount of space that each plant and animal takes up. For example, you might count one grasshopper, two young trees and estimate that half or 50% of the square is covered with grass.
6. Draw a sketch of each plot to show the numbers of plants and animals and the way they are spread out in the square.
7. Observe and record the changes that occur in the plots each week for several months. Compare the information you have collected from your plots with other groups.
8. Write a summary of the changes that have occurred during your project.

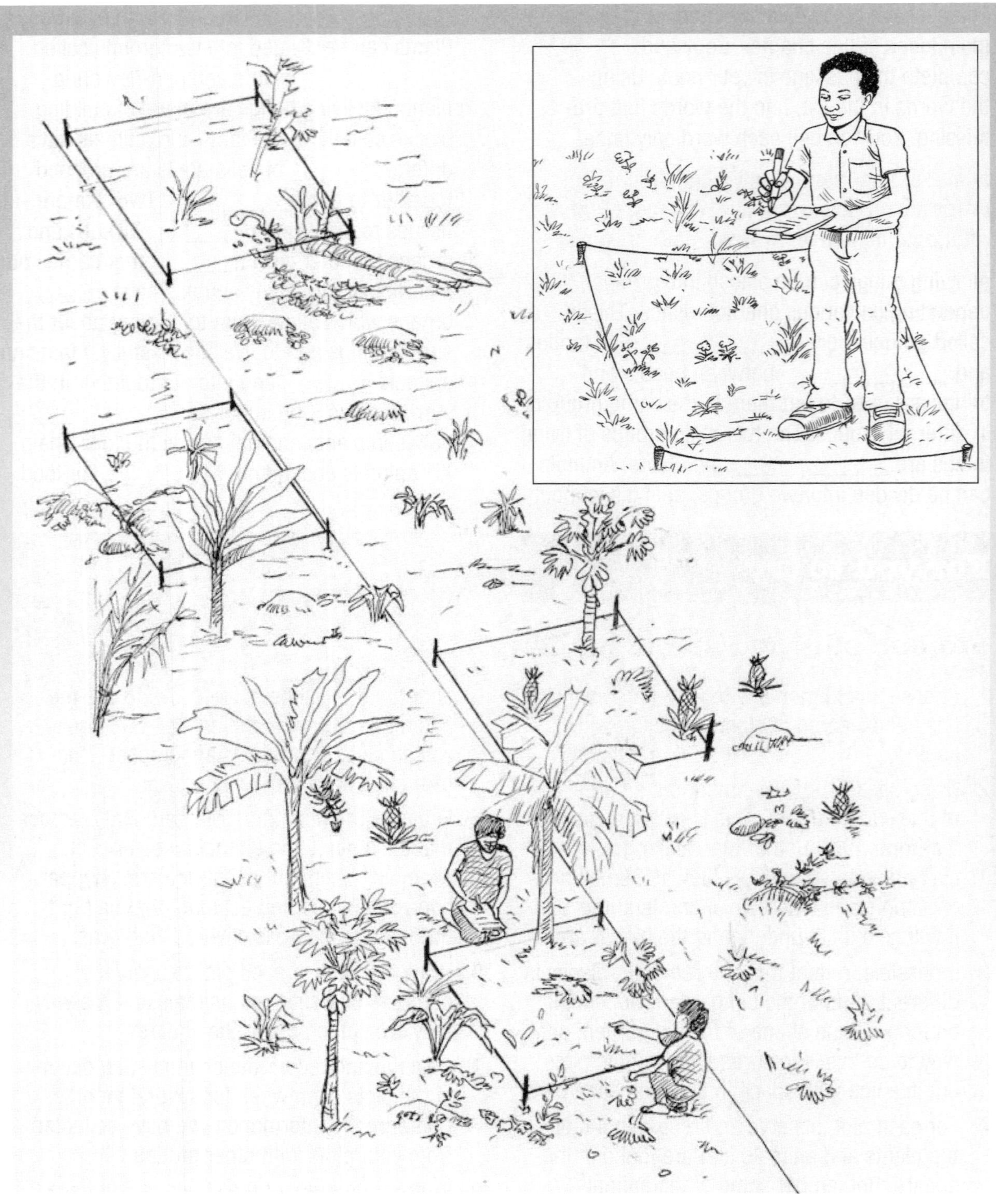

Studying an old garden

Project: Making a model ecosystem

1 Collect a large, empty plastic soft drink bottle, a sharp knife, some good soil, some leaf litter, a few small plants, and a few small animals, e.g. an earthworm, a snail, some slaters.

2 Cut the plastic bottle into three pieces about the same size. You do not need the middle piece.

3 Fill the bottom of the bottle with good damp soil and plant one or two healthy small plants.

4 Put one or two earth worms in the soil and cover with damp leaf litter.

5 Put the snail and the slaters on the leaf litter.

6 Fit the top part of the bottle over the base. Leave the cap screwed on.

7 Place your model ecosystem where it will get light, but not in direct sunlight.

8 Observe and record what happens for several weeks. Describe what happens.

9 Your model ecosystem should keep going for a long time without watering. How can you explain this?

10 You can add water if needed and you should check it from time to time to make sure that the animals have not died.

11 You may also need to remove the lid for one hour, once a week.

12 Different groups of students could try using different plants and different animals to find out which are the most suitable.

3

Science in the home

Chapter summary

In this chapter you will have an opportunity to:

- find out about common materials or substances that are found in the home and community
- find out about mixtures of substances and be able to share this information with others
- find out about different sources and types of energy in the home
- find out about and be able to explain how simple machines can be used in homes and the community to do work.

Syllabus references

Strand: Science in the home

Sub-strand: Learning about substances

Outcomes:

6.3.1 Identify and organise common substances into groups according to physical properties

6.3.2 Conduct practical investigations into the nature of mixtures and communicate their findings in a scientific way

6.3.3 Identify and describe the sources and the types of energy

6.3.4 Identify and describe the nature of force as being a push or a pull

6.3.5 Identify and explain how simple machines can be used in homes and the community to do work

Key facts

- All substances can be classified into groups known as solids, liquids and gases depending on their properties.
- Some substances can mix together and some cannot mix together.
- A mixture is formed when one substance dissolves in a liquid to form a solution.
- A substance that is dissolved in a solution can be separated from the liquid.
- A substance that will not dissolve in a liquid may form a suspension.
- Some mixtures of substances that do not dissolve can be separated by decanting.
- Energy is the ability to do work.
- Energy can be changed from one form to another.
- Different sources of energy are used in the home.
- Simple machines can be used to do work in the home and the community.
- There are five different types of simple machine: levers, pulleys, axles, inclined planes and gears.
- A force can be a push or a pull.
- Forces make objects move.
- Friction is needed to allow things to move, but friction will also make things slow down or stop.
- Gravity is a force.

Learning about substances

There are many things that we use every day that we find around the home or at school. We wear clothes, we sleep in a cotton bedsheet, we eat food that comes from a tin and drink from a plastic cup or a plastic bottle. We may have a roof made of corrugated iron, a floor made of concrete and louvres made from glass. We might play with a rubber ball or go fishing with plastic line and a lead weight.

There are similarities and differences between all things depending on what they are made of. For example, shirts and skirts are often made of a mixture of polyester and cotton. Food tins and corrugated iron are both made from iron. However, tins that contain food have a thin coating of tin, and corrugated iron has a thin coating of zinc.

Scientists describe all of these things as **substances**.

Things that we use every day are all examples of substances.

The nature of matter

Scientists say that all things are made of **matter**. Matter is anything that has **mass**. In other words matter is something that can be weighed and takes up space. Mass is measured in grams and kilograms. There are 1000 grams in one kilogram and 1000 kilograms in one tonne. We use scales or balances to measure mass.

Scales and balances are used to measure mass.

People are made of matter and take up space.

All the things that you use every day, the things you eat and all the objects around you are made of matter. The air that you breathe is also made of matter and has mass. You are made of matter and take up space. Two people cannot be in the same space at the same time.

There are thousands of different kinds of matter in the world. Some of these are natural materials like wood, rock and water. Some kinds of matter are made by people, like plastic, glass, paper and cordial.

Certain types of matter, such as iron, copper, lead, oxygen and mercury, are simple substances called **elements**. There are over 100 different elements. Other types of matter, such as plastic, paper, brass, water and wood, are made up of a combination of different elements. Substances that are made by joining two or more elements together are called **compounds**. For example, sugar is a compound of the elements carbon, hydrogen and oxygen and salt is a compound of the elements sodium and chlorine.

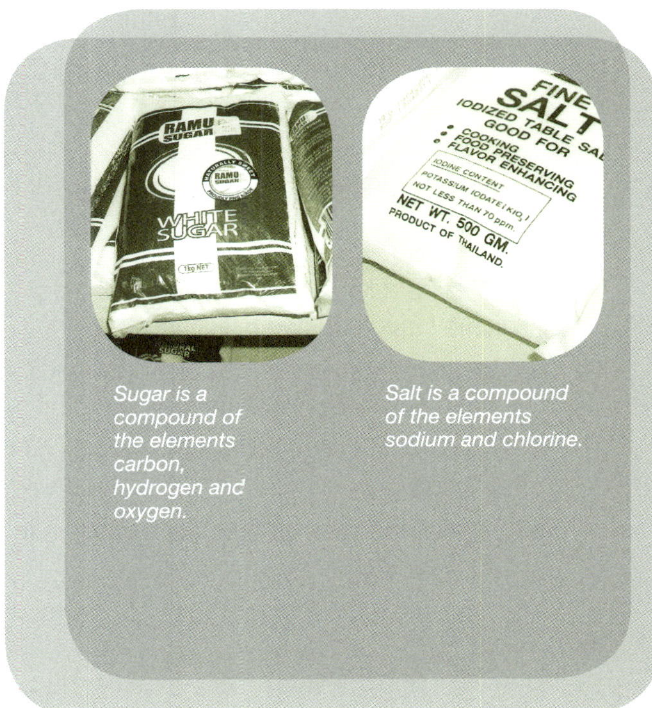

Sugar is a compound of the elements carbon, hydrogen and oxygen.

Salt is a compound of the elements sodium and chlorine.

For you to try

1 Make a list of the things that you have used today and say whether they are natural materials or made by people.

2 Look at the labels of some different kinds of clothes, food or drink to find out what they are made of or what ingredients they contain.
   - a What are the most common materials used to make clothes?
   - b Which of these are natural and which are man-made?
   - c What are the most common ingredients in the food and drink?

3 Look at as many different substances as you can find at home and at school and copy and complete the following table:

Substance	Description	What is it used for?
Kerosene	Blue coloured liquid	Cooking and lighting, refrigerators

4 Investigation: Matter has mass

- a Collect a ruler, a long thin straight stick, some string and two balloons.
- b Put the ruler over the end of a desk and hold it down with some books.
- c Blow up the two balloons and tie each one to the stick.
- d Balance the stick on the ruler as shown in the diagram.
- e Burst one of the balloons, but be careful not to move the stick – you may need to gently hold the stick while you burst the balloon.
- f Observe what happens and explain what you observe.
- g In this investigation, the matter that you use is air. Does air have mass?

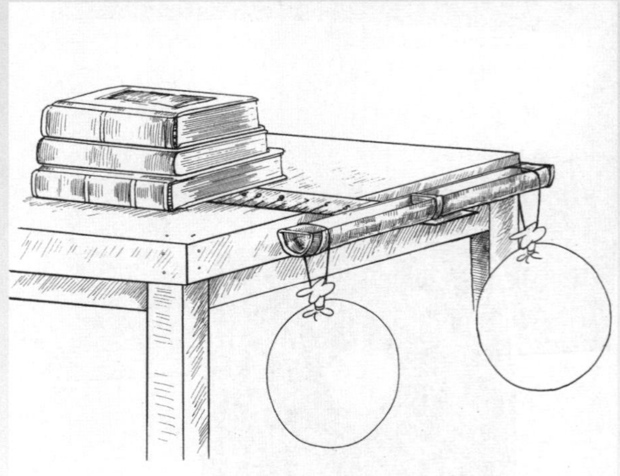

5 Investigation: Matter takes up space

- a Collect two empty plastic drink or cordial bottles and a knife with a sharp point or a hammer and nail.
- b Screw the top on one of the bottles tight. Squeeze the bottle several times. What happens? As you squeeze the bottle undo the cap. What happens?
- c With a sharp pointed knife carefully make a small hole in the plastic top of one of the bottles. You can do this by gently turning

the knife into the plastic or using the hammer and nail.

- **d** Cut the bottom off this bottle as shown in the diagram. Cut it close to the bottom so that it will make a funnel that can hold a bottle of water when it is upside down.
- **e** Remove the cap from the other bottle.
- **f** Fill your funnel with water and carefully balance it over the empty bottle.
- **g** Make sure that the outside of the cap and the lip of the empty bottle are wet to make a seal and stop air getting out of the empty bottle (you must position it carefully and may need to push down gently if the seal does not form).
- **h** Observe what happens and explain what you observe.
- **i** Discuss what happened in this investigation with other students in your group.

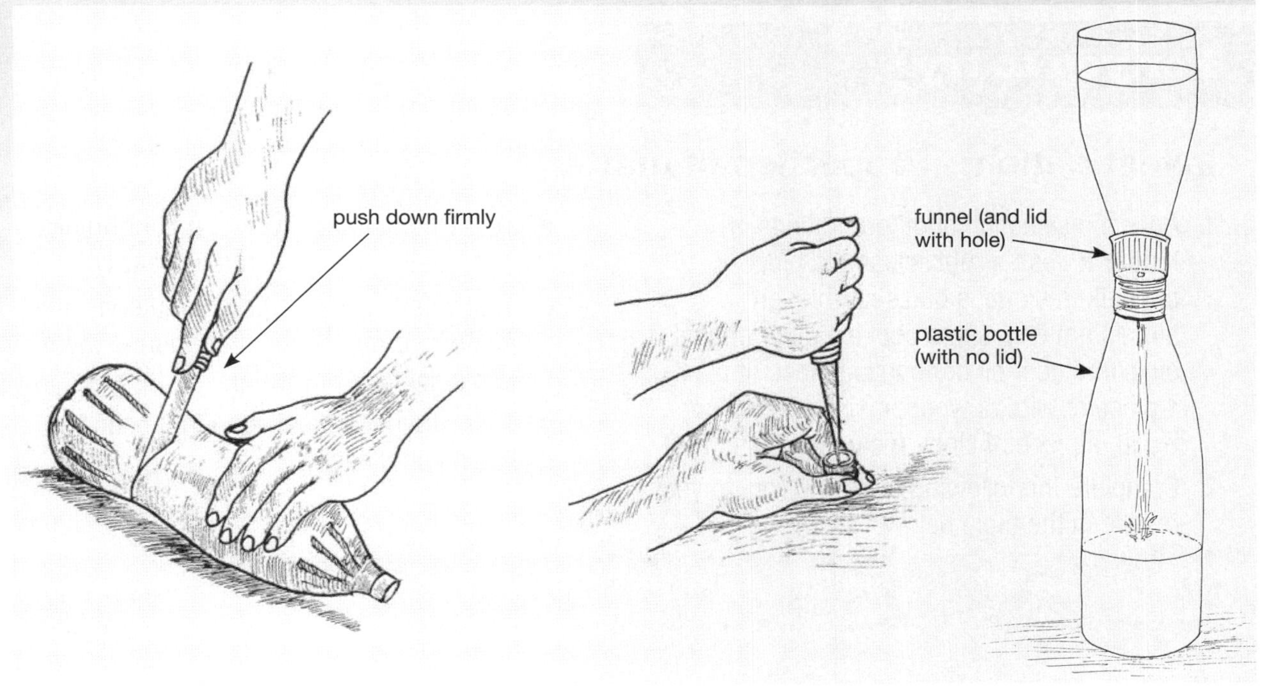

The properties of matter

People use different materials for different purposes. We usually choose the type of material because of the **properties** of the material. For example, we might choose to use a clay pot for cooking because it makes the food taste good, but an aluminium saucepan because it will not break so easily and is easier to carry. We use a lead sinker because it will carry our fishing line to the bottom of the sea. Elastic is used in the waistband of shorts and skirts because it will stretch each time to fit a small person or a large person.

The table below includes words that you can use to help describe the properties of matter and some examples of materials that have this property.

Materials and the properties of matter

Property	Yes	No
Breaks easily?	Weak, e.g. clay pot, glass	Strong, e.g. steel, wood
Bends easily?	Flexible, e.g. rubber thongs, sheet of flat iron	Stiff, e.g. handle of tools.
Stretches again and again?	Elastic, e.g. rubber band, rubber ball, waistband of clothes	Non-elastic, e.g. string, rope
Pours easily?	Fluid, e.g. water, kerosene	Viscous, e.g. gear oil, honey
Sinks easily in water?	Dense, e.g. lead sinker, stone	Light, e.g. polystyrene, some wood, pumice
Scratches easily?	Scratchable, e.g. most plastics	Hard, e.g. steel, some rocks, diamond

For you to try

Investigation: Properties of matter

SR

1 Collect a selection of different kinds of matter to test: a rubber band, a piece of chalk, a stone, a glass marble, an aluminium can, some cooking oil or engine oil, some gear oil or honey, a piece of plastic, a piece of wire, a wooden stick, a block of wood, a piece of cloth, a candle, etc.

2 Complete the following tests and record your results in the table below.

a Test whether each type of matter breaks, bends, stretches or scratches easily.

b Test whether each type of matter sinks or floats by dropping them or pouring them into a container of water.

c Test whether a small amount of each liquid pours easily.

Type of matter	Breaks easily?	Bends easily?	Stretches again and again?	Scratches easily?	Sinks or floats?	Pours easily?

3 Write some sentences to answer the following:

a Which type of matter broke most easily?

b Which type of matter bends most easily?

c Which types of matter stretched easily and then went back to their original shape?

d Which types of matter were most easily scratched?

e Which type of matter did not float?

f Which liquid was the most difficult to pour?

The three states of matter

All matter can be divided into three groups called **solids**, **liquids** and **gases**. These are known as the **states of matter**.

1 Solids

Solids are substances like wood, stone, iron and plastic. Solids have a particular shape and they take up space. Therefore solids also have a fixed volume. Solids are also rigid which means that the shape will not change unless some force changes them.

We can pick up a solid to move it and it will not change its shape. For example, a steel nail or a marble will not change its shape unless the nail is bent by a hammer or the marble is crushed.

Solids cannot be squashed or compressed into a smaller volume.

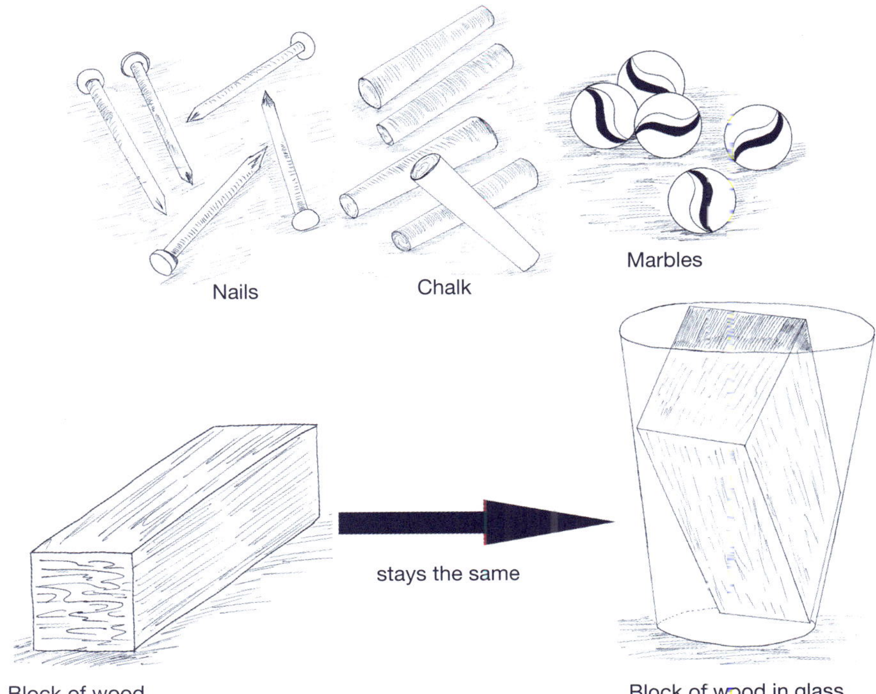

Examples of solids

2 Liquids

Liquids are materials like water, kerosene, blood, cordial and mercury.

Liquids have no shape of their own and are able to flow. This means that they can be poured like water out of a cup or flow like water in a stream. Liquids are not rigid and always take the shape of the container that holds them.

Like solids, a given amount of liquid has a fixed volume. Although the liquid will take the shape of the container, its volume stays the same.

Liquids cannot be squashed or compressed into a smaller volume. This is a very useful property which means that a liquid can be pumped and used to do work in a machine. For example, the brakes on a car, hydraulic jacks and the three point linkage on a tractor work by pumping special oil which moves along a pipe but does not change its volume.

When a large number of small solid particles, such as grains of sand, are put together they can behave like a liquid. You can see this if you pour a cupful of sugar or dry sand into a different container. The sugar or sand takes the shape of the container.

The sand timer makes use of the fact that particles of solid sand can behave like a liquid. The sand falls from one side of the timer to another and it can later be turned over and the sand falls the other way. The sand timer used in the National Parliament measures two minutes before a vote is taken.

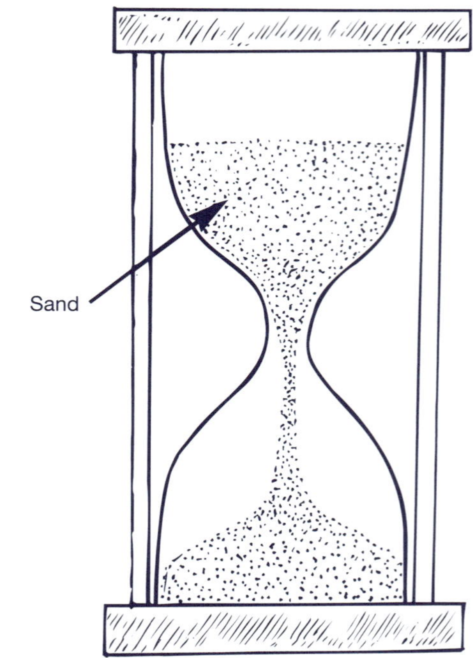

A sand timer

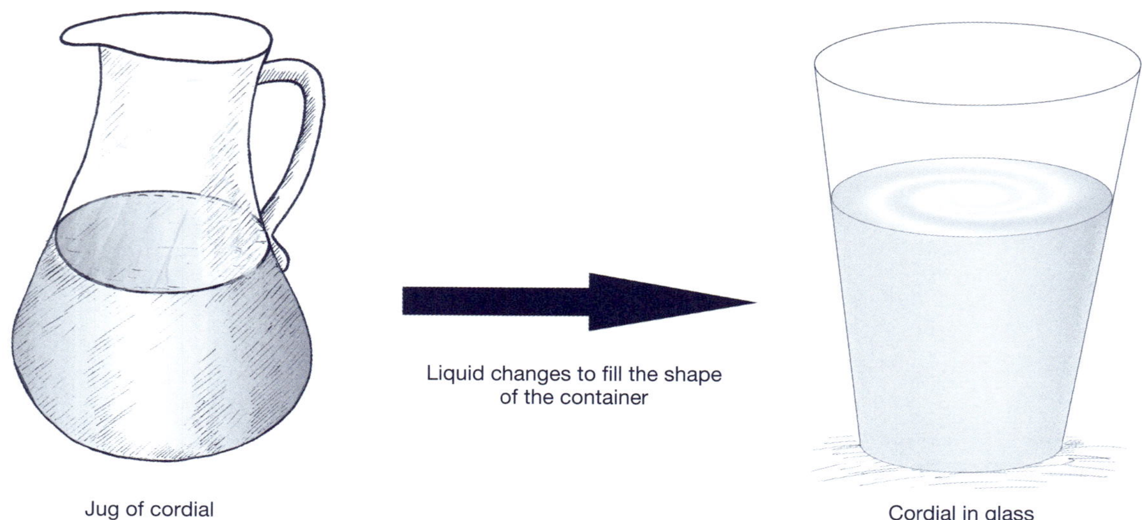

Examples of liquids

For you to try

Investigation: Making a sand timer

1. Collect two plastic drink or cordial bottles, sticky tape, a strip of cloth, strips of rubber from an old inner tube, or string to tie the bottles together.
2. You will also need a knife with a sharp point or a hammer and nail, and a watch or clock so that you can compare your sand timer with the watch or clock.
3. With a sharp pointed knife or the hammer and nail carefully make a small hole about 3mm in diameter exactly in the centre of the plastic tops of both bottles. You can do this by gently turning the knife into the plastic. Both holes must be in the same position.
4. Put some fine dry sand into one of the bottles and replace the cap. Measure and record the amount of sand using a cup or scoop. This will allow you to change the amount later.
5. Fix the two bottles together so that the holes in the tops are together. You will need to use sticky tape, a piece of cloth, strips of rubber from an old inner tube, or string.
6. Measure the time taken for the sand to fall from one bottle to the other.
7. Change the amount of sand in the timer or the size of the hole, so that it falls for a whole number of minutes, for example, three, four or five minutes. You can put marks on the outside of the bottle to show different lengths of time. This is called calibrating your timer.
8. Show your timer to other groups and notice different ways that people made their sand timers.
9. Make a list of the problems that you had in making and calibrating your sand timer and the way that you overcame those problems. Share your list with other groups.
10. Use your timer for some activities. For example, how many times can two people throw a ball to each other while the sand timer is running?

3 Gases

Gases are different from solids or liquids because they have no fixed shape or volume. When a gas is put into a container it will spread throughout the container and will take the shape of the container. For example, if there is a gas leak or if something is rotten inside a house you can smell it in different places. This is because the gas spreads out to all parts of the room. When something spreads out like a gas we say that it **diffuses**.

Gases can be squashed or compressed into a smaller volume if they are held inside a container. This property can be tested with a syringe. If you put your finger over the outlet of the syringe and push the plunger in, the air can be squeezed into a smaller space. When the plunger is let go, the air pushes the plunger out and the air returns to its original volume. You can do the same thing with a bicycle pump or a pump used for sports balls.

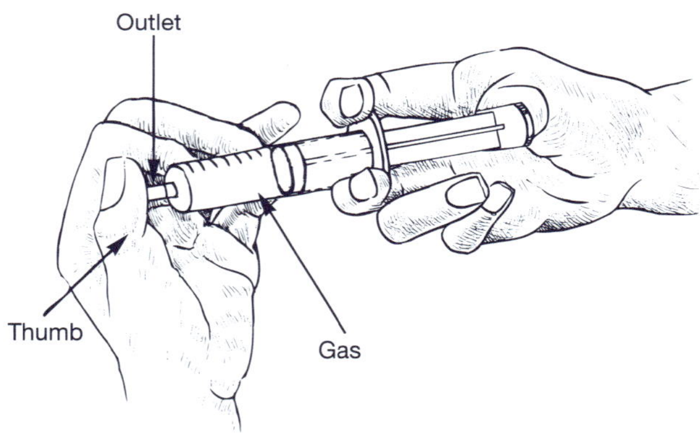

Gases can be compressed into a smaller volume.

The air returns to its original volume because of its **elastic property**. In the same way, a foam mattress gets thinner when you lie on it but will return to its original shape when you get up.

The tyres of bicycles, motorbikes, cars and trucks also make use of the elastic property of air. The air in the tyres helps to cushion the shocks from the bumps in the road. Other examples which make use of the elastic property of air are balls such as netballs, basketballs, soccer balls and footballs, which all have air pumped into them.

Petrol and diesel engines also work because the mixture of air and fuel can be compressed into a

Tyres filled with air cushion the shocks from bumps in the road.

When a ball is bounced or kicked we are using the elastic property of air.

smaller space and then expands rapidly when it is ignited and burned inside the engine.

Because of its elastic property, a large amount of air can be compressed into a small metal tank which can be carried on a person's back and used for breathing. For example, firemen carry tanks to allow them to work in smoke-filled buildings for about half an hour. Scuba divers can stay under water for as long as an hour on a single tank of air. Scuba stands for self contained underwater breathing apparatus.

Petrol
Sparkplug
Petrol
Sparkplug

As the fireman breathes from the tank the compressed air returns to its original volume.

A diver using self contained underwater breathing apparatus (scuba).

Gas for cooking comes in bottles.

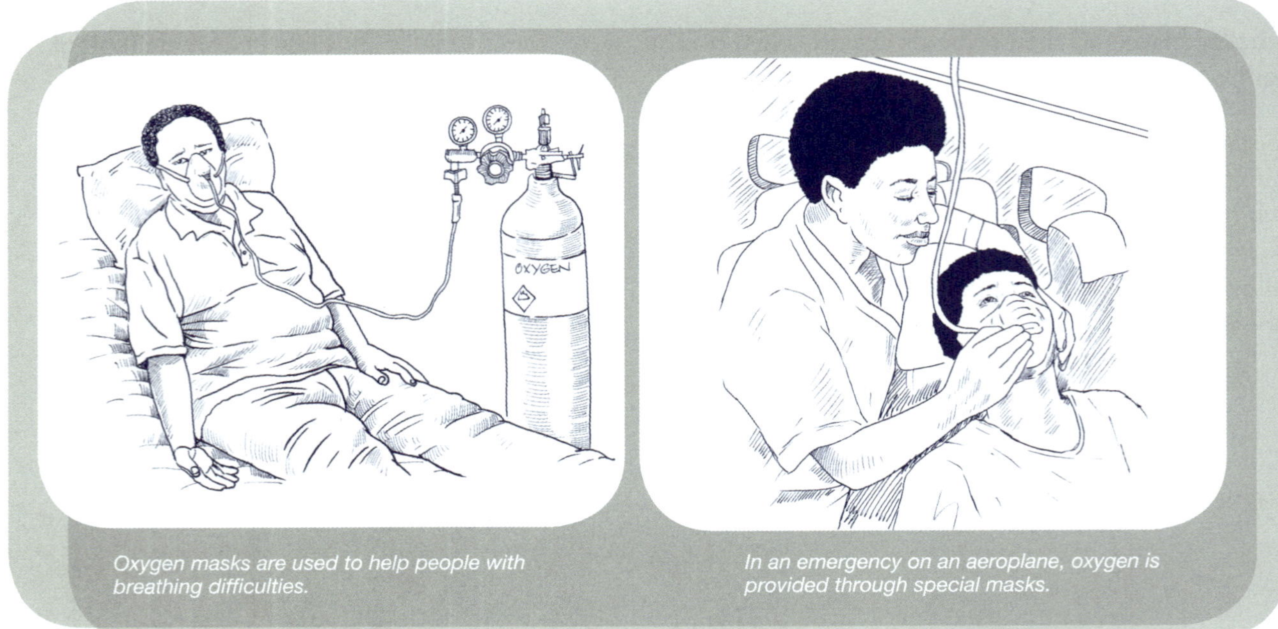

Oxygen masks are used to help people with breathing difficulties.

In an emergency on an aeroplane, oxygen is provided through special masks.

Gas for cooking which is compressed into bottles is also used in many places and hospitals also use oxygen in bottles to help patients who find it hard to breathe. Large aeroplanes also carry oxygen for the passengers, but this is only used in an emergency.

Carbon dioxide is the gas produced by living things as a waste substance and is also used by plants to make food. Humans also use carbon dioxide. For example, carbon dioxide is used in fizzy drinks to make the bubbles. If carbon dioxide is poured onto a lighted candle, the flame will go out. For this reason carbon dioxide is used in some fire extinguishers.

Carbon dioxide gas provides the bubbles in fizzy drinks.

Carbon dioxide gas makes a good fire extinguisher.

For you to try

1 Investigation: Finding out more about gases

SR

a Collect a bucket or dish and two glasses or cups.

b Fill the bucket or dish with water.

c Push the glass mouth down into the water.

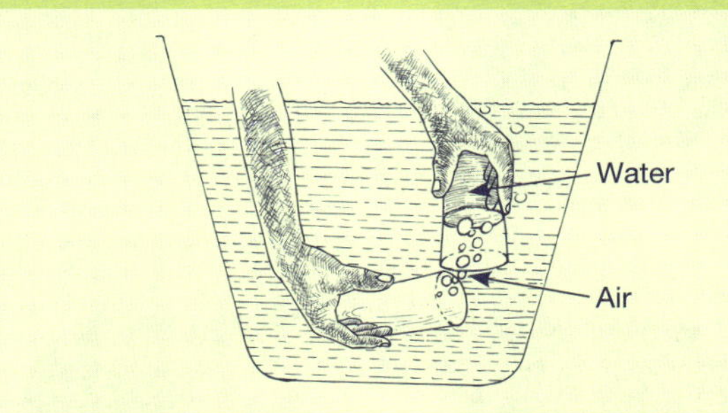

Air is a gas and takes up space.

Take care—the glass will be slippery.

d Observe what happens to the air inside the glass. What can you see and feel?

e Put the second glass in the bucket, let it fill with water and hold it upside down above the glass containing the air.

f Tilt the first glass gently to allow the air to escape slowly and try to catch it in the second glass.

g Observe what happens.

h What conclusions can you make about air?

i Describe your investigation to say what you did and what you found out.

2 Copy and complete the following table to show the common things that are found in your home or school, putting each in the correct column depending on its state. A

Solids	Liquids	Gases

What can you say about the number of things in each group? Can you give a reason for this?

Change of state

The state of matter depends on the **temperature** and this means that matter can change from one state to another. For example:

- when a candle is burning some of the wax will melt and run down the side of the candle, and then it will become solid again
- an ice cream or an ice block will melt quickly on a hot day
- wet clothes get dry when they are hung outside

A change of state can occur in both directions. For example:

solid → liquid → gas or gas → liquid → solid

Heat is needed to make the first set of changes happen, and **cooling** must occur to make the second set of changes happen. We use special words to describe these changes of state and these are shown in the table below.

Change of state	Words used to describe the change	Heat or cool to make the change?
solid → liquid	melting	heat
liquid → gas	evaporation	heat
gas → liquid	condensation	cool
liquid → solid	solidification or freezing	cool

Many substances around us can exist in different states, depending on the temperature. For example, water is a common substance that can exist in all three states. When water is cooled to 0 degrees Celsius a change of state occurs. Water changes to ice, which is a solid, and we say that the water has **frozen**. When water is heated to 100 degrees Celsius, it boils and another change of state occurs. The water changes to steam, which is a gas, although we cannot see steam. We say that the water has **evaporated**.

For you to try

Investigation: What happens when a candle burns?

- **a** Collect a candle and some matches.
- **b** Place the candle where you can observe it safely and easily.
- **c** Light the candle and observe carefully what happens to the candle wax. In particular, look closely at the changes of state that occur.
- **d** Describe and explain your observations. Use a labelled diagram to help you.
- **e** Can you think of any similar examples where the same changes of state occur due to heating and cooling?

Making use of the change of state

Most plants and animals need water in order to be able to live. Rain provides fresh water which runs into rivers and the sea and which can then be used by plants and animals, including humans. Heat from the sun causes water to evaporate, which eventually forms clouds and falls again as rain. This is called the **water cycle** and so we depend on the fact that water can easily change state from liquid to a gas and back to a liquid.

Solder can easily be melted and used to join metals together. For example, plumbers use solder to make water tanks and gutters which are very important in many parts of Papua New Guinea. Electricians also use solder to join wires together.

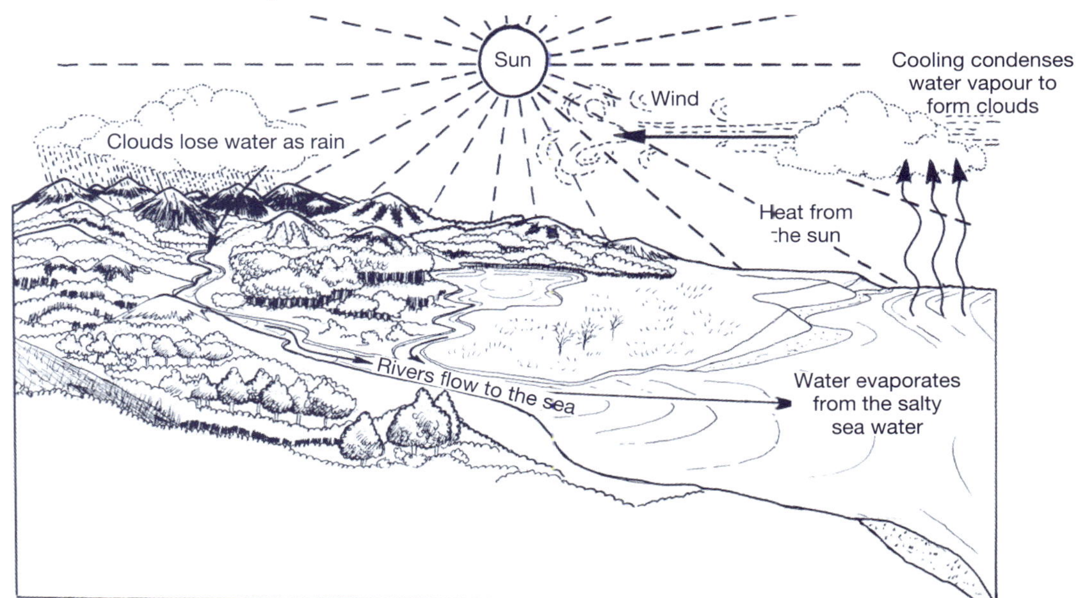

The water cycle

Rainwater tanks and gutters are made by soldering galvanised iron.

Copper pipes can be joined by soldering.

Some metals are heated until they are liquid and then they are poured into moulds to make things like fishing sinkers, taps, hammers, axes and parts of engines. Pouring liquid metal into a mould to get the shape you want is also called making a **cast** or **casting**. Some coastal people collect the lead from old car batteries and heat it over a fire and then pour it into a mould to make fishing sinkers. The skin from a betel nut makes a good mould for casting a fishing sinker.

Most plastic bottles, bowls, cups, other containers and toys are made by injecting liquid plastic into a mould. If you look carefully you can usually see a small rough spot where the plastic came out of the mould.

When aluminium cans are recycled in Port Moresby they are melted in a furnace and the aluminium is poured into moulds.

For you to try

1 Using your own knowledge and information from this chapter, copy and complete the following table to summarise the properties of solids, liquids and gases.

State	Shape—fixed or takes shape of container?	Volume—fixed or fills volume of container?	Able to be compressed?
Solid			
Liquid			
Gas			

2 Explain why the tyre of a bicycle gives a smoother ride when it is filled with air than when the tyre is flat.

3 Look at different plastic objects to find the spot where the plastic was injected into the mould. It is about the size of the point of a pencil and feels a little sharp. Look for similarities and differences in the position of this spot. What can you say about the position of this spot for most of the plastic objects that you look at?

4 Copy and complete the following table which describes some common task or events. One example has been completed for you.

Task or event	Change of state	Word describing the change
Filling a petrol tank and noticing a smell	Liquid to gas	Evaporation
Making iceblocks		
Heating water to make a pot of tea		
Soldering		
Leaving butter out of the fridge		
Making a fishing sinker		

Mixtures and solutions

One substance by itself is called a **pure** substance. It might be an element such as gold or it might be a compound such as water. However, very few natural substances are pure. For example, rainwater is almost pure but contains tiny amounts of other chemicals as well. The black charcoal found after a fire is carbon that is almost pure.

If something contains at least two separate substances, it is called a **mixture**. Most of the things around us are mixtures. For example, air is a mixture of gases, mainly nitrogen and oxygen. Sea water is a mixture of salt, water and other compounds.

Mixtures can be made by mixing substances together that are in different states.

Charcoal is carbon that is almost pure.

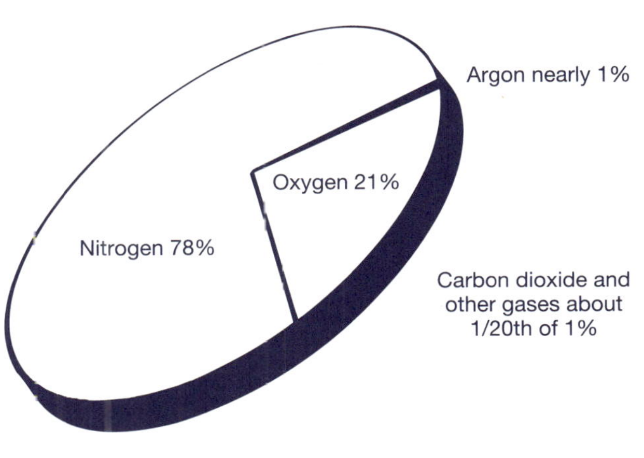

Air is a mixture of gases.

The formation of mixtures

States being mixed	Example
Solid mixed with solid	Sand mixed with nails
Solid mixed with liquid	Sand mixed with water
Liquid mixed with liquid	Water mixed with cordial
Liquid mixed with gas	Water mixed with oxygen
Gas mixed with solid	Air mixed with soil
Gas mixed with gas	Nitrogen mixed with oxygen in air

Living things are even more complicated. Plants and animals are mixtures of many thousands of compounds mixed together in very special ways.

Face painting uses traditional mixtures.

Important traditional mixtures

PD

People in Papua New Guinea have known for a long time how to make their own paints and dyes and these mixtures are used for painting the body and for dyeing grass skirts, bilums and baskets. Some of these colours come from plants, using the bark, leaves, stem and sap. Some colours come from special soils like ochre and clay and from special rocks. For example:

- vegetable pigments—yellow, red-brown, scarlet, blue and green
- burning special wood—red
- clay—white, orange, brown
- ochre—buff (light brown), yellow, orange
- powdered soapstone—grey

For you to try

Investigation: Traditional paints and dyes

A

- **a** Collect examples of traditional paints and dyes in your area.
- **b** Identify the source of the materials used to make the paint or dye.
- **c** Describe how they are mixed and how they are used.
- **d** Make a poster or colour chart to show examples of these traditional paints and dyes.

Metals

Metals are useful because of their properties. Most metals are hard solids that can be flattened into sheets and stretched into wire. For example, aluminium is used in cooking pots, copper in wires and lead in car batteries. Sometimes two metals are joined together because of their properties. For example, most food cans are made of iron covered with a layer of tin. Sheets of iron coated with a layer of zinc are used as galvanised iron. In both cases the iron is protected from rusting.

Copper is a good electrical conductor and can be shaped into wire

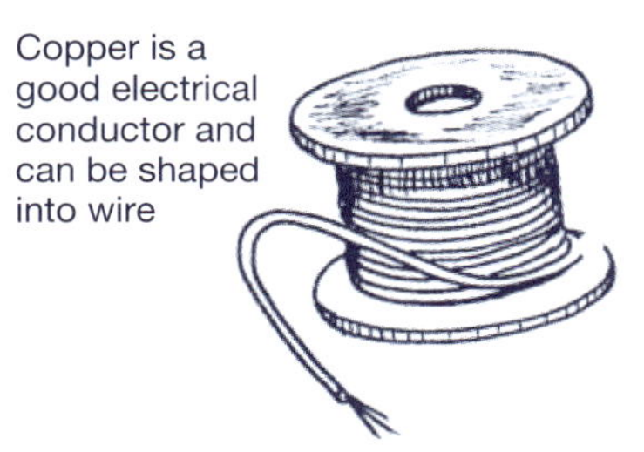

Zinc is used in the case of dry cells and takes part in the chemical reaction that produces electricity

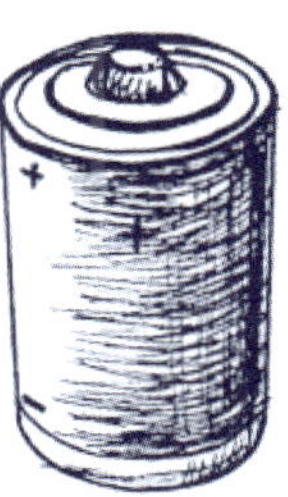

Special carbon steels are used for cutting tools because they keep their cutting edge

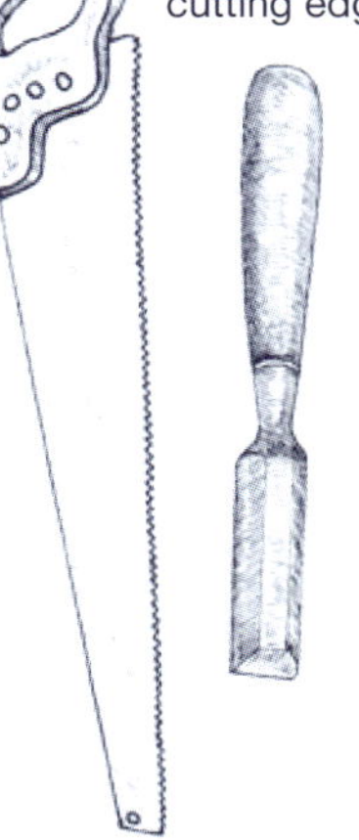

Manganese steel is used to make the jaws of a bulldozer because it is very tough and hard-wearing

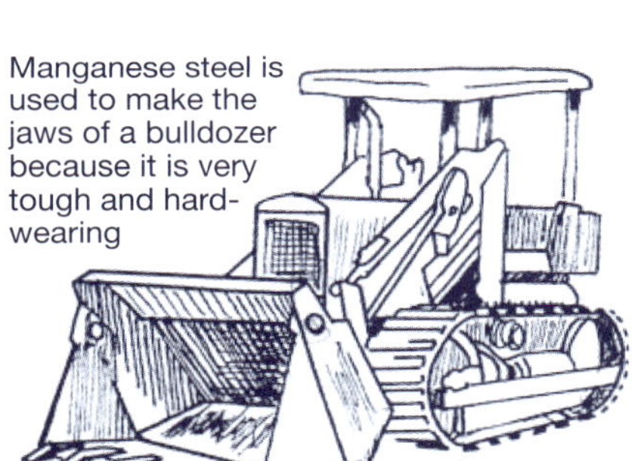

Hardened steel is used to make hammer heads

Aluminium is one of the few metals that can be made into a thin sheet called foil

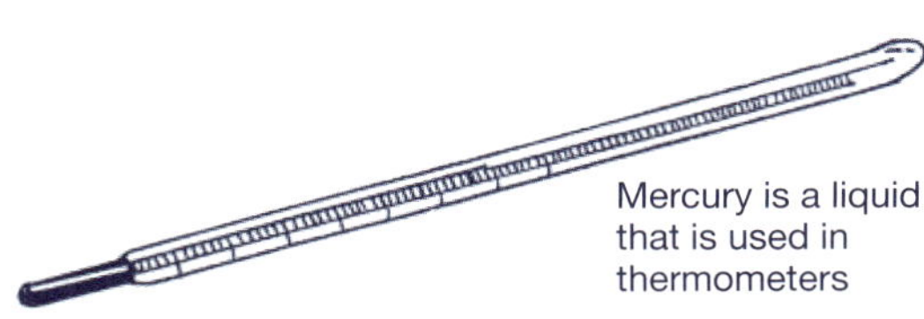

Mercury is a liquid that is used in thermometers

Metals are useful because of their properties.

For you to try

Make list of the metal objects that you have used today. Give the name of the object and the metal from which it is made.

Solutions

When sugar or salt is added to water it **dissolves** or breaks into very small particles that are too small to be seen. The tiny particles spread throughout the water which is clear. You cannot see the salt or the sugar but you can still taste it.

When a substance dissolves in a liquid we say that a **solution** is formed. A solution is a mixture of substances. When something will dissolve in water we say that it is **soluble** in water.

The substance that does the dissolving is called the **solvent** and the substance that dissolves is called the **solute**. Some solutions may be coloured but all solutions are clear, which means that you can always see through them and there is no sign of cloudiness.

Water is a good solvent and is easily available and this is why we can use water to wash our clothes and make them clean again. We use water to make the dirt on our clothes dissolve and the washing powder helps this process to work more easily. We also have to move the clothes about in the water to help the dirt to dissolve, just like stirring a cup of tea helps the sugar to dissolve. After washing our clothes, the water looks dirty

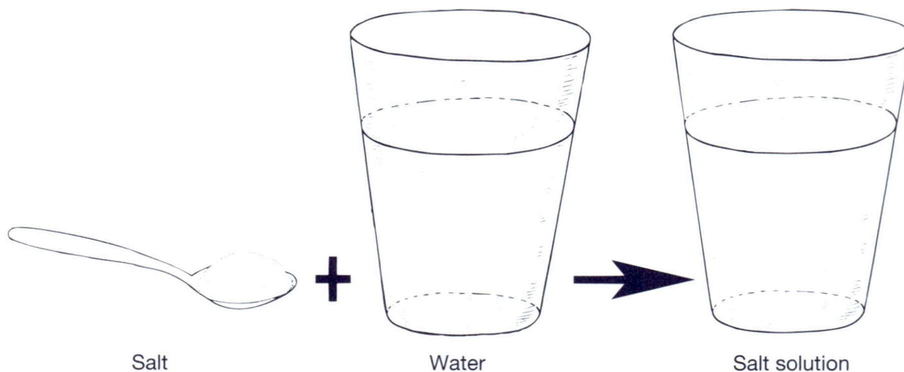

All solutions are clear with no sign of cloudiness.

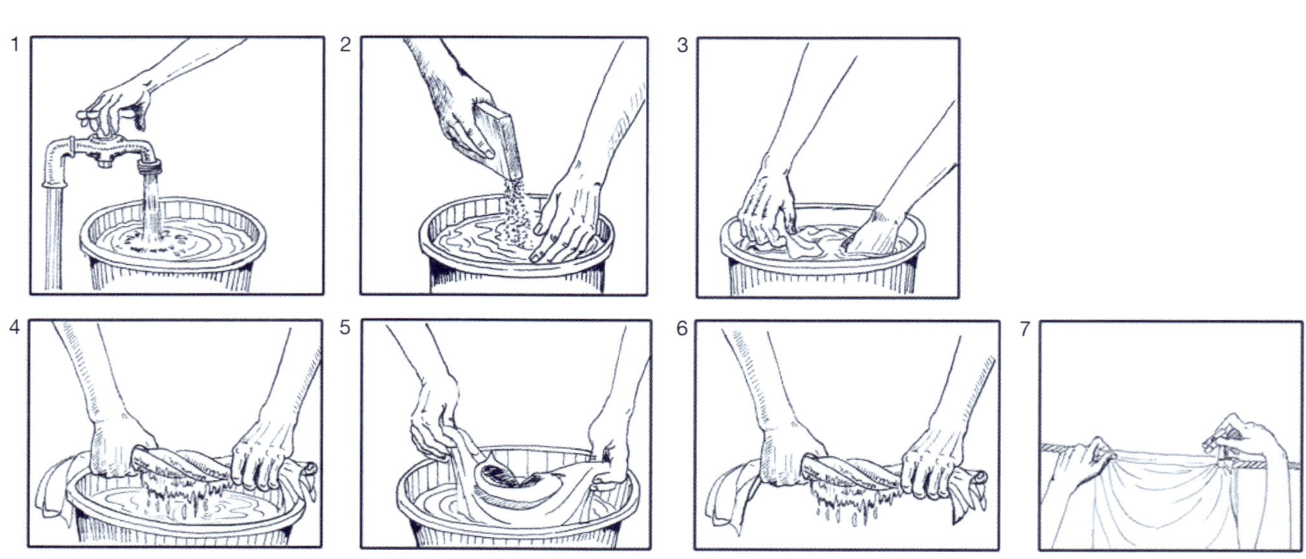

Steps in washing clothes
1. Put water in sink, bowl or bucket. 2. Add soap powder and clothes. 3. Move clothes about in water, rub.
4. Wring or squeeze. 5. Rinse clothes in fresh water. 6. Wring or squeeze. 7. Hang out to dry.

because the dirt has dissolved in the water. We usually wring the clothes or squeeze them to remove the soapy water. Rinsing in fresh water helps to remove the last of the dirty, soapy water from the clothes.

However, not everything will dissolve in water and so other solvents must be used. For example:

- kerosene and petrol will dissolve oil and grease
- methylated spirits will dissolve the ink from a biro
- turpentine will dissolve oil-based paint, so is added to paint that is too thick and is also used to clean brushes
- acetone (nail polish remover) will dissolve nail polish.

Turpentine is a good solvent and is used to thin oil-based paint and clean paint brushes

For you to try

1 Investigation: Does it dissolve? SR

a Collect a small amount of different substances like sugar, salt, flour, milk powder, instant coffee, curry powder, jelly crystals, oil and soil etc. Also collect a glass jar, a teaspoon and water.

b Half-fill a glass jar with clean water.

c Add half a teaspoonful of one of the substances and stir.

d Observe what happens and record your result in the table.

e Clean the jar and repeat with the other substances.

Substance mixed with water	Observations
	(e.g. does it dissolve — clear or cloudy? colour?)

f Classify the substances that you have investigated according to whether they dissolve in water.

2 Make a list of common solutions that are found around your home or school.

Insoluble substances

Sugar and salt both dissolve in water and we say that they are **soluble**. However, sand and many other substances do not dissolve in water and we say that they are **insoluble**. When we add sand to water we can clearly see that it is a mixture.

We can prove that it is a mixture by separating the sand and water.

If chalk dust is shaken in water the tiny pieces of chalk will spread throughout the water making the liquid look milky. However, the liquid will not be clear. This is called a **suspension** because the particles are suspended in the water. Muddy water is another example of a suspension. When a suspension is allowed to stand for a long time the little pieces of undissolved substance slowly settle to leave a **sediment** on the bottom.

Some medicines, sauces and cleaning materials are made as suspensions, which is why the directions on the bottle say 'Shake well before use'.

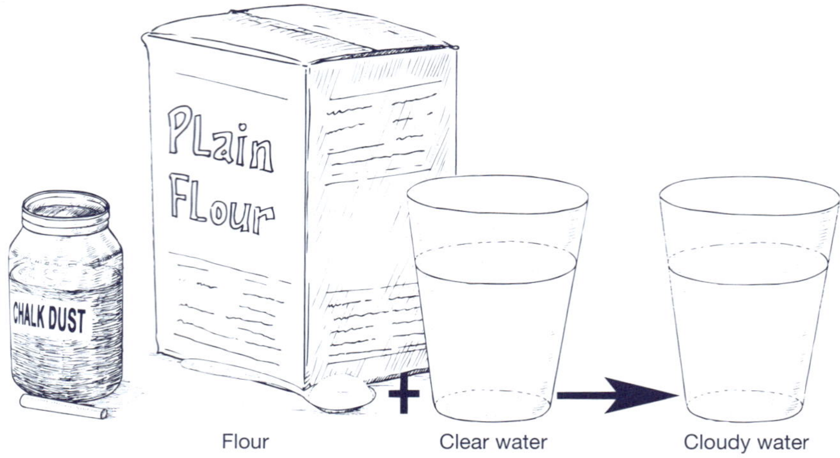

Flour and chalk do not dissolve in water but form a suspension.

For you to try

Investigation: Soil and water

a Collect a small amount of soil and add to some clean water in a clear glass jar or clear plastic container.

b Shake well and allow it to stand for several days.

c Observe and describe what happens over this time. Draw a diagram with labels to help you explain your observations.

Separating mixtures

There are two ways to separate an insoluble solid from a liquid.

Decantation

If the solid is heavy it will sink to the bottom of the liquid and the liquid can be poured off carefully leaving the solid behind. For example, some people wash rice before cooking in order to get rid of any insects or other dirt. They add water to the rice and then stir it with their hands. Then they carefully and slowly pour the water away leaving the clean rice in the bottom of the pot. This process is known as **decantation** or decanting.

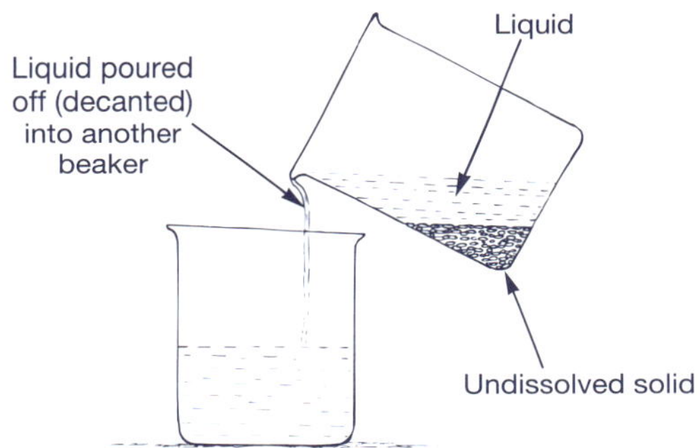

Separating an undissolved solid from a liquid by decantation.

Filters used in a car engine.

Filtration

In muddy water the insoluble material is fine and light and does not sink to the bottom quickly. The solid can be separated by pouring the cloudy mixture through a **filter** which allows the liquid to pass through but catches the solid. This process is known as **filtration** or filtering. Many kinds of filters like sieves and strainers are used in the home for preparing food. Filters are found in car and truck engines. Traditional fish traps, fishing nets, fly wire on the windows of houses and mosquito nets are also kinds of filters. Sometimes we use the material that passes through the filter, and sometimes we use the material that does not pass through the filter.

Scientists use a special kind of paper called a **filter paper** to separate a solid from a liquid. The liquid passes through tiny holes in the paper, but the solid collects on the paper. Filter paper usually comes in the shape of a circle and is folded into quarters and then opened up to make a cone shape which is then placed in a funnel. You can make your own filter paper using old newspaper and you can make your own funnel by cutting the top off a plastic drink or cordial bottle.

Filters used at home.

Filter paper folded to fit in a funnel.

Living things also make use of filters. For example, the hairs in the nose filter dirt from the air that is breathed in, so that the air reaching your lungs is clean. Some animals are able to obtain their food by filtering the water in which they live. These animals are called **filter feeders**. Shellfish with a double shell and whales feed in this way.

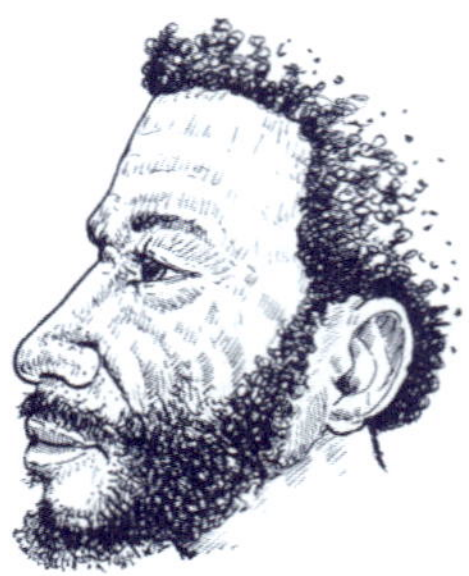

a The nose is a filter.

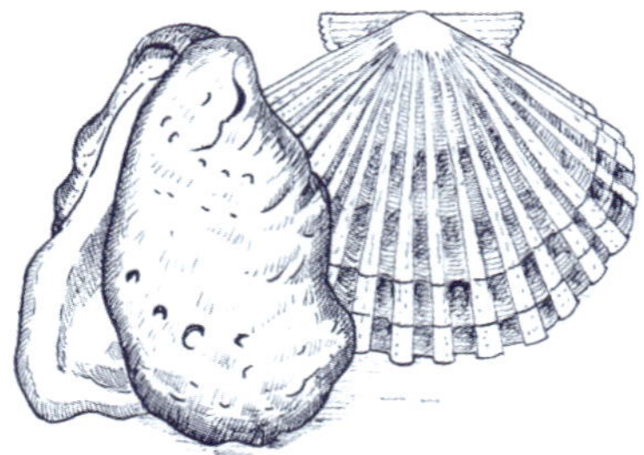

b Many shellfish are filter feeders.

c Whales are filter feeders.

For you to try

1 Investigation: Finding out about mixtures

a Collect a plastic bottle, six containers, a sheet of old newspaper, a cotton handkerchief (or other material), a piece of fly wire, sand, flour (wheat flour or sago flour), chalk powder or lime for chewing with betel nut, talcum powder, mud and some water.

b Make the following mixtures: sand and water, flour and water, crushed chalk powder or lime and water, talcum powder and water, mud and water.

c Keep each one in a separate container. Stir each mixture well before you test it.

d Try decanting each mixture by pouring off the liquid and leaving the solid behind. Record your result.

e Make your own funnel and container by cutting the top off a plastic bottle.

f Make three different types of filter: one from a sheet of old newspaper, one from a cotton handkerchief (or other material) and one from a piece of fly wire.

g Pour each mixture through each of the filters and observe what happens. Compare each of the three filters.

h You will have to use a new piece of paper each time and wash the others before you test each mixture.

i Record your observations using a table of results like the one on the next page.

Method of filtering

Mixture	Decanting	Paper filter	Material filter	Flywire filter
Sand and water				

a Put the filters in order from the best to the worst.

b What will happen if you repeat the experiment with a salt solution and a sugar solution?

c Test your prediction and record your observations.

d Suggest an explanation for your observations.

2 Copy and complete the following table to show the different kinds of filters that are used in your home or in your area. One example has been done for you. A

Example	What is the filter made of?	What things are being separated or filtered? (say which state they are in)	What is the purpose?
Tea strainer	Plastic or metal	Tea leaves (solid) and tea (liquid)	Stop tea leaves from going in the cup

When you have completed your list, divide them into two groups: those in which we use the things that have passed through the filter, and those in which we use the things that have not passed through the filter.

Science in the village: Making sago

The sago palm grows in many parts of Papua New Guinea and many years ago people discovered how to get sago flour from this plant. For this reason, sago is an important staple food that is eaten in many places. Sago flour is almost pure starch and is an important source of carbohydrate. Because starch is an insoluble substance it can be separated from water. The following steps describe the method used to make sago.

1 The sago palm is cut down, the trunk is cut in half lengthways and the pulp inside is broken into very small pieces. The pulp is a mixture of plant fibre and flour.
2 The pulp is placed in the stem of a palm frond on the river bank. Water from the river is used to wash the pulp.
3 As the water passes through the pulp, the small grains of flour are separated from the fibre and flow with the water through a **filter** made from a piece of coconut hair (or flywire).
4 The tiny particles of insoluble sago flour and water pass through the filter, but not the coarse insoluble sago fibres.
5 The flour and water mixture flows into a large container, and the flour settles to the bottom. The water overflows without losing any flour.
6 Washing the pulp continues until all the flour has been extracted. The water is then poured off, or **decanted**, leaving the wet flour in the container.
7 The wet sago flour is wrapped in leaves and hung over a fire to **evaporate** the water and dry the flour.

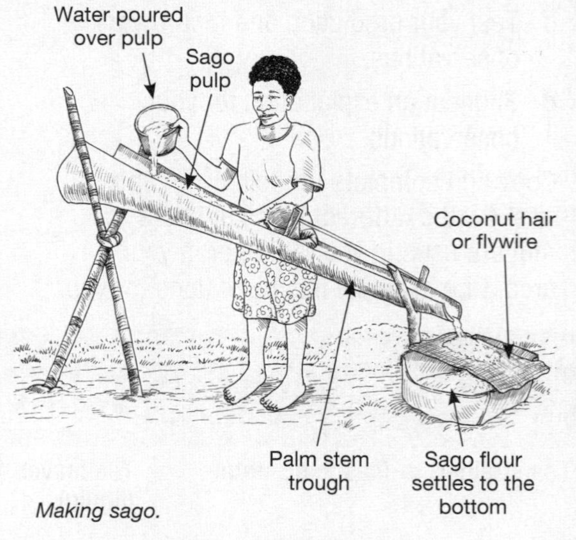

Making sago.

Separation by evaporation

When one substance is dissolved in another we cannot separate them by filtration. Another way of saying this is that we cannot separate a solute from a solvent by filtering.

To separate a soluble substance like salt or sugar from water, we use a process called **evaporation** because the salt or sugar is dissolved in the water. In evaporation the solution is heated. The heat causes the water to evaporate into the air, leaving the solid behind. For example, if we leave a dish of salty water in the sun the water will evaporate, leaving behind particles of white salt. In order to make the process of evaporation quicker we can heat the solution on a stove or with a fire.

Getting salt from our surroundings

On the coast we can get salt from sea water by using the process of evaporation. For example, at LeaLea salt lakes near Port Moresby, a wide rim of salt crystals forms at the edge of the water. The salt forms slowly as the water evaporates due to the heat of the sun. This salt can be collected and used for cooking. Salt is collected in this way in Australia and other parts of the world. It takes about two years for the sea water to pass through a series of ponds before the salt can be harvested.

In the highlands we can get salt from small pools or lakes that contain quite large amounts of salt. For example, people from the Laiagam area of Enga province have a traditional way of getting salt.

- Logs from soft wood trees are put into the pool to soak up the salt water.
- After about six months the logs are taken out and left to dry in the sun.
- The logs are then burnt in a very hot fire.
- When the fire has died away, the fine ash is blown away leaving hard grey lumps which are a mixture of white salt and unburnt carbon.
- These lumps are collected, wrapped in leaves and stored in a dry place.

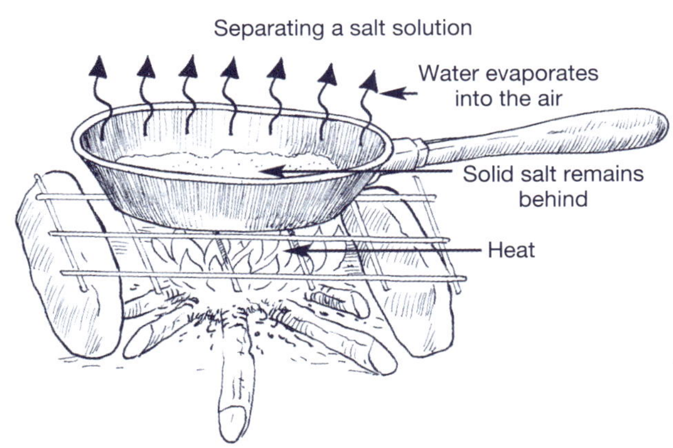

Getting salt from salt water.

We can get salt from sea water.

- The salt parcels were used for cooking and could also be used for trading with the Mendi people for their crude oil.

Purification of salt

The salt that we get from burning logs that have been soaked in salt water is mixed with grey ash. The salt can be purified by separating it from the ash as follows:

- Water is added to the mixture of salt and ash.
- The salt dissolves in the water but the ash does not dissolve.
- The mixture is heated to make sure all the salt dissolves
- The mixture is filtered using fibre obtained from a plant. The salt solution passes though the filter but the ash remains behind on the filter.
- The salt is obtained from the solution by heating.
- When all the water has evaporated, the white salt crystals are collected in the dish.

Traditional Salt

1 Logs from soft wood trees are put into a pool to soak up the salt water

2 After six months, logs are taken out

3 They are left to dry in the sun

4 The logs are then burnt in a very hot fire

5 When the fire has died away, and the ash is blown away, hard grey lumps are left which are a mixture of white salt and unburnt carbon

6 Lumps are collected, wrapped in leaves and stored in a dry place

Cooking

7 Salt parcels can be used for cooking

Barter system

8 Salt parcels can be traded with others, for example, the Mendi people for their crude oil

For you to try

1 Investigation: Getting salt from sea water

- **a** Collect a small amount of sea water and place it in a shallow dish (if you do not live near the sea you can make your own sea water by adding salt to water).
- **b** Leave the dish in the sun so that the water can evaporate.
- **c** When the water evaporates add a little more sea water to the dish.
- **d** Keeping adding more sea water for as long as you can.
- **e** Describe what happens.
- **f** How long does it take before you can see salt in the dish?
- **g** How do you know that it is salt?
- **h** How could you make salt more easily?

2 Draw a series of diagrams to explain how sago flour is made in the village. Show clearly the different stages where filtration, decantation and evaporation are used. Remember to label your diagrams.

3 How could you remove the following substances from drinking water?

a sand

b sawdust

Draw diagrams to explain your answer. Remember to label your diagrams.

Magnets

Substances that contain iron or steel are attracted to a magnet. This means that a magnet can be used to separate iron or steel when it is mixed with something that is not magnetic. For example, when car tyres are recycled an electromagnet is used to separate the steel wire from the rubber. An electromagnet can also be used to separate the iron and steel from metals that do not contain iron.

An electromagnetic crane can be used to separate iron and steel from metals that do not contain iron.

For you to try

Investigation: Separation by magnetism

a Collect a small piece of steel wool or scourer that is used for cleaning saucepans, a pair of scissors, some sand, a magnet and a sheet of paper or plastic.

b Cut up a small amount of steel wool or scourer into small pieces.

c Mix the pieces of steel wool or scourer with a little dry sand.

d Use the magnet to separate the mixture of steel and sand.

e Repeat the experiment but try to find a way to stop the pieces of steel wool from sticking to the magnet. (Hint – which piece of equipment has not been used yet?)

f Describe your experiment to say what you did and what you found out.

Science in the village: Getting fresh water using a solar still

Sometimes people get lost in the bush or in a canoe or small boat and may not have enough food or water. People need food but water is more important. In hot places people can survive for only about three days without water. When you cannot find fresh water, it is possible to make fresh water using a **solar still**. A solar still uses the heat of the sun to make water evaporate from plants, from the ground or from the sea. The solar still then collects the water when it condenses back to a liquid. This method has saved lives when people get lost in boats or when they are walking in the bush, especially in dry places.

1. Collect a plastic sheet, a container like a cup or tin and some stones.
2. Choose a place where there is a natural hollow or a slope in the ground.
3. Dig a hole in the ground at the lowest point and place a container like a cup in the hole.
4. Cover the area with a plastic sheet like a tarpaulin.
5. Place a stone above the cup to hold the plastic down at this point. This must be the lowest point of the plastic sheet or tarpaulin.
6. Place other stones around the edge of the plastic sheet.
7. The edges of the plastic must be higher than the cup so that the water can run along the underside of the sheet towards the cup.
8. Taste some of the water that collects in the container. If you have a straw you can taste the water without taking the still apart.
   - Does it taste salty?
   - Does it taste like tap water?

11 How long does it take before you can get enough water to drink?

12 Make a poster to show how your solar still works. Label your poster and use the words 'evaporation' and 'condensation'.

13 The process in which a liquid evaporates to a gas and then condenses back to a liquid is called **distillation**. The name solar still comes from the process of distillation that the still uses.

14 How can you use this method to get fresh water from salty water? If you live near the sea, try out your ideas.

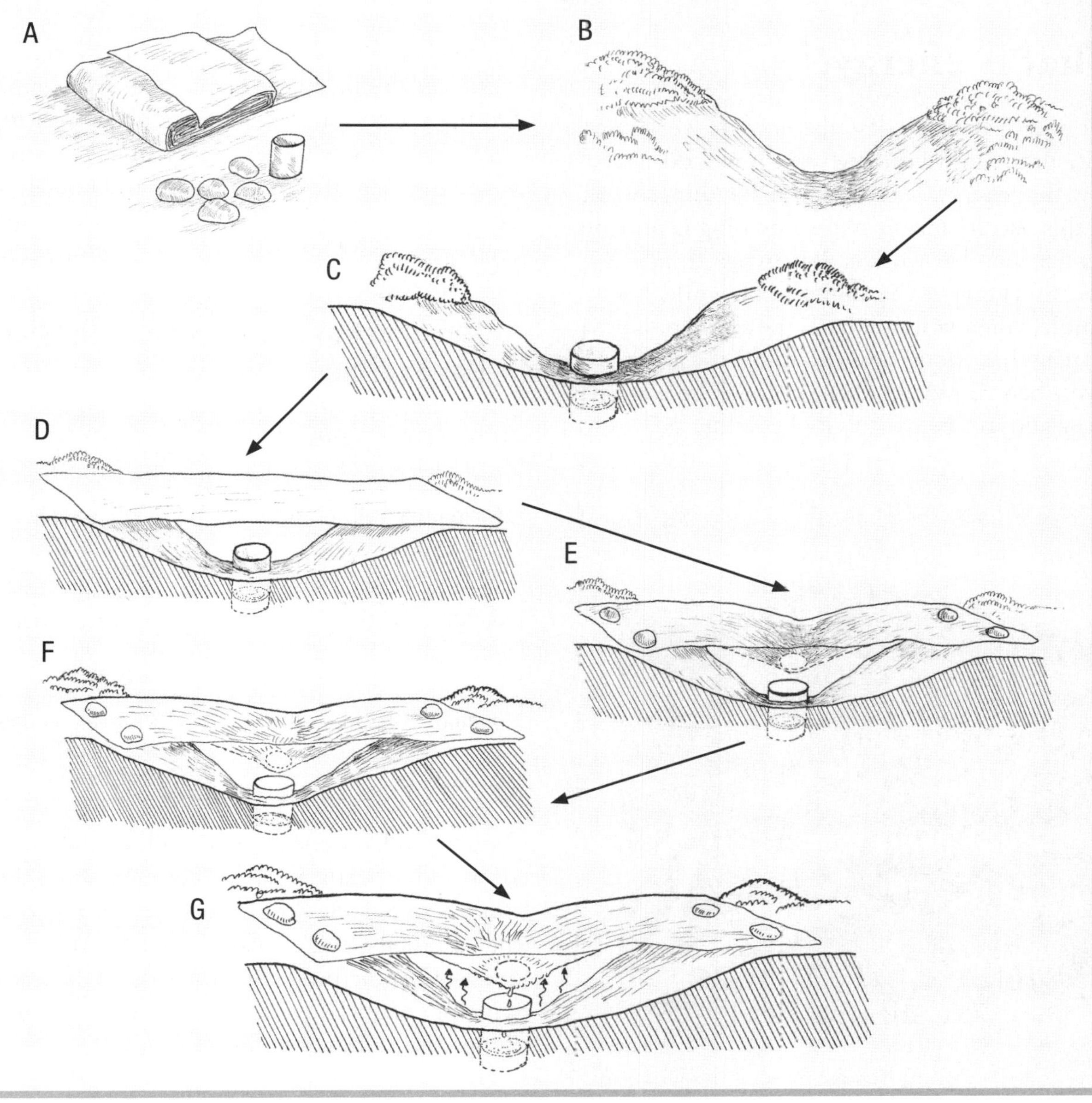

Using energy in the home

What is energy?

Energy is difficult to describe because we cannot see it, although we can see what it will do. We know that energy is needed to make things work and this usually means that something is moving. Scientists say that work is done when things move. So energy is the ability to do work. For example, when you pull back the string of a bow, the string has energy and can make the arrow move when you let go of the string. The wind has energy and can be used to make sailing canoes and yachts move across the water. However, wind can also be strong enough to cause damage. During the cyclone season from November to April, Papua New Guinea sometimes gets strong winds that can destroy buildings and knock down trees. The table shows the effect of some cyclones in the Louisiade Archipelago of Milne Bay Province:

The energy of a cyclone can destroy buildings and knock down trees.
(Photograph of Tropical Cyclone Larry © The Australian Bureau of Meteorology)

The effect of some cyclones in Milne Bay Province

Date	Name of cyclone	Damage caused
November 1967	Annie	Some people killed, boats destroyed
January 1988	Agi	Wharves, houses and crops destroyed
January 1989	Aivu	Wharves, houses and roads destroyed

People also have energy which allows them to do work and, of course, people get tired when they run out of energy. People make things move as they do their work or play games. Anything that can make something else move has energy. Other types of energy allow us to cook our food, to see things and to hear sounds.

Types of energy

Different types of energy can be summarised as follows:

Type of energy	What is it? What does it do?	Examples
Heat energy	Cooks our food, keeps us warm	The sun, fire, our bodies, heating metals to shape them, engines
Light energy	Lets us see things	The sun, the moon, kerosene lamp, candles, torch, light bulbs
Sound energy	Lets us hear things. Sound is produced when something vibrates.	Voices, drums, wind in the trees, moving water, animals, aeroplanes
Electrical energy	Can be used to make sound, make things hot, make things cool, make things move	Lightning, batteries, generators, power stations
Moving energy (kinetic energy)	When something is moving it has energy.	Wind, falling water, bicycles, trucks, ships
Stored energy (potential energy)	Energy inside a substance that can be used when it is changed to another type. Energy because of the position of something.	Food, kerosene, petrol, batteries Water in a dam, coconut in a tree

Using energy

Energy is needed to transport people from place to place. Energy is used in homes, offices and factories to run all kinds of machines. Energy is used for heating and lighting and for cooking and storing food.

People need energy to do things.

Cars and trucks get energy from burning petrol or diesel in the engine. The energy that powers a radio is supplied by the batteries. When you paddle a canoe or ride a bicycle the energy comes from the muscles in your body. Your muscles get the energy from the food you eat. When you cook food on a fire the energy comes from the firewood. Electrical energy is used to operate lights, cooking stoves, washing machines, radios and televisions, petrol pumps and so on.

In the examples given we have seen that energy can do the following things.

- Energy will make things move.
- Energy will make things hot.
- Energy will make sound.
- Energy will make things happen.

In all of these examples energy is changed from one form to another.

For you to try

A

1 Copy and complete the following table to show the things or devices that we use every day and that need energy to do work. The first example has been completed for you.

Thing or device	What is it used for?	Where does the energy come from?
Radio cassette player	Making sounds, listening to music	Batteries

2 Find out about different types of machines that can be used to generate electricity. Make a poster to show one of these machines and use labels to help explain how it works.

3 Unscramble each of the following words that have been used in the previous paragraphs.

a Krow
b Gynere
c Rowpe
d Snuod
e Tighl

Main sources of energy

The sun is our most important source of energy because almost all the energy we use comes from the sun. This energy is called **solar energy** and travels through the air as a special kind of wave. When solar energy reaches the earth it heats up the air, water and ground.

Plants are able to use energy from the sun to make their own food by a process called **photosynthesis**. They use water from the soil and carbon dioxide from the air to make sugars. These sugars are later changed to more complicated substances in the plant. The process of photosynthesis can be summarised as follows:

$$\text{carbon dioxide} + \text{water} \xrightarrow{\text{solar energy}} \text{oxygen} + \text{plant food (sugar)}$$

The plant stores the energy that came from the sun. Later on the plant will use this stored energy to grow. When we eat parts of plants like fruits and vegetables, seeds and nuts, we are eating the parts in which the plant has stored food. When we eat food that comes from animals, like eggs, fish and pork, we are also using energy from the sun because chickens, fish and pigs also get their energy from eating plants.

Energy is stored in food.

Other sources of energy are:

- firewood from trees
- fossil fuels like coal, oil and natural gas that are found underground
- running water
- wind.

Firewood

Firewood is an important source of energy in all parts of Papua New Guinea. Many people collect firewood and some people buy and sell firewood. Firewood is used to cook food, to keep warm and to give light. In some places firewood is used to dry crops. For example, people on the coast use firewood to dry the meat from coconuts when they make copra. We are able to use this energy because trees are able to trap the energy of the sun and store it in the form of wood. Because firewood comes from trees that can be replaced we say that firewood is a **renewable energy resource**. In some parts of PNG people plant special trees in order to get firewood. For example, in the highlands people plant casuarina trees for firewood.

Fossil fuels

The energy from the sun that reached the earth millions of years ago was trapped by plants and changed into forests and jungles. When dead plants rot and decay in the ground over millions of years they can form substances like coal, crude

The energy from the sun is stored in coal.

Collecting and decanting oil in a bamboo.

oil and natural gas. Coal looks like a black rock and is not normally found in PNG. However, PNG is lucky to have oil and gas under the ground. Traditionally, people in the Southern Highlands have collected and used the oil where it seeps to the surface of the ground. Mining companies have also drilled for oil in Papua New Guinea and this is now being taken out of the ground and used to make petrol and diesel fuel. There are now also plans to build a long pipeline to send gas from Papua New Guinea to Australia. The gas will be used to make electricity and this is expected to start in 2009. Selling oil and gas to other countries is a way for Papua New Guinea to earn money.

Coal, oil and natural gas are known as **fossil fuels**. Fossils are the remains of plants and animals that died a long time ago. Fuels are substances that can be burned to give energy.

Fossil fuels are the most important source of energy used in the world today. However, they cannot be replaced and one day they will run out. In other words, fossil fuels are a **non-renewable energy resource**.

Running water

Water is another natural source of energy. Running water can be used to turn machines. The water is usually stored in a dam or reservoir in the mountains and can be used to produce electricity. The water runs down large pipes into a power station. The water gains a lot of energy as it falls down the mountain and is used to spin a turbine, which is a special propeller that is connected to a generator. As the generator turns it produces electricity called **hydroelectricity.**

Papua New Guinea is lucky enough to have high rainfall and fast flowing rivers in the mountains and so is a good place to make hydroelectricity. Almost half of the electricity produced in Papua New Guinea is hydroelectricity, so it is an important source of energy. The main hydroelectric schemes are described on the next page, but there are also small schemes in many places.

- The Rouna power stations get their water from the Sirinumu dam in the mountains near Sogeri and supply Port Moresby with electricity. The same water is also used for the water supply of Port Moresby.

- The Ramu scheme at Yonki, near Kainantu in the Eastern Highlands, supplies electricity to Kainantu, Goroka, Kundiawa, Mount Hagen, Lae and Madang.
- The Warangoi in East New Britain supplies power to the Gazelle peninsula.

A hydro electric generator.

The Rouna hydroelectric scheme.

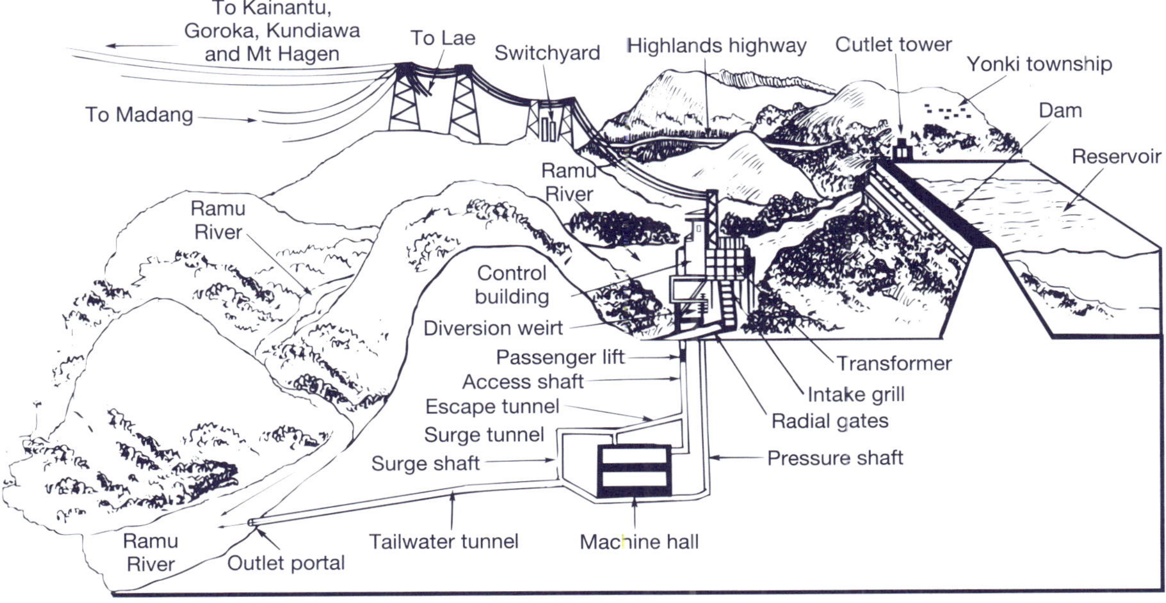

The Yonki hydroelectric scheme.

Hydroelectricity is a renewable resource but requires a good supply of water. It also does not produce waste or pollution.

Wind

Wind is a natural source of energy that was traditionally important in Papua New Guinea. For example, in the Hiri trade between Central and Gulf provinces, the people depended on the wind for their sailing canoes, or lakatois. The people of the islands of Milne Bay still make sailing canoes and sailing boats and canoe races are held around Port Moresby.

In some countries wind farms have become more important in recent years. Wind farms usually have many big propellers and are used to generate electricity.

A sailing canoe

A wind farm

For you to try

1 Investigation: Different types of energy

a Collect a square piece of paper or thin card, a pin, a stick, a torch, a candle and matches, and a bilum with something heavy inside, like sweet potato or a pile of books.

b Observe what happens in each of the following experiments and describe what happens:

i Make a paper windmill using the square piece of paper, pin and stick. Blow on the windmill or hold it outside in the wind.

ii Rub your hands together very quickly.

iii Turn on the torch.

iv Light the candle. Blow it out when you have finished.

v Clap your hands together.
vi Put the weight in the bilum and hang it on a hook so that it can swing easily and give it a push to make it start moving. When something swings backwards and forwards it is called a pendulum.

c Copy and complete the following table to show what you did and your observations (what did you see, hear and feel):

d Describe the different types of energy that you observed.

e Discuss your results with other groups.

Experiment	What did you do?	What happened?	Type of energy
i			
ii			
iii			
iv			
v			
vi			

2 Investigation: Making a wheelie-reel

a Collect an empty cotton reel, a strong rubber band, a matchstick, a pencil or stick, sticky tape, and a metal washer.

b If you cannot find a cotton reel you could use a short piece of bamboo or make your own using two plastic lids and a short piece of plastic pipe.

c Thread a strong rubber band through the hole in a cotton reel.

d Put a matchstick through the loop in one end and tape the matchstick to the end of the cotton reel so that it will not move (or cut a little groove for the matchstick).

e Put a metal washer over the other end of the rubber band and then put a pencil through the rubber band.

f Wind up the pencil until the rubber band is tightly twisted.

g Put the wheelie-reel on the floor and let it go.

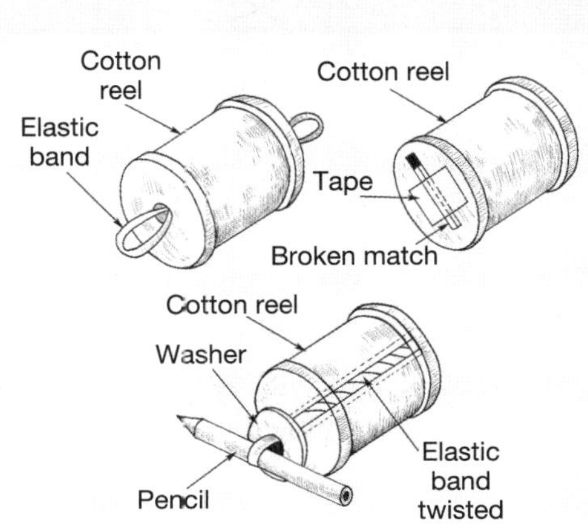

- **h** Answer the following questions about your wheelie-reel:
  - **i** How far did your wheelie-reel go?
  - **ii** What is the source of energy for your wheelie-reel?
  - **iii** When the rubber band is wound up, what sort of energy does it have?
  - **iv** Energy is needed to wind up the rubber band. Where did this energy come from?

3 Describe the sources of energy in your area and the way in which they are used in everyday life.

4 Use the graph shown below to answer the following questions:

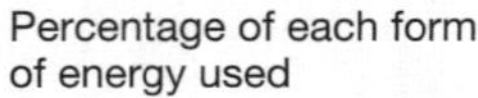

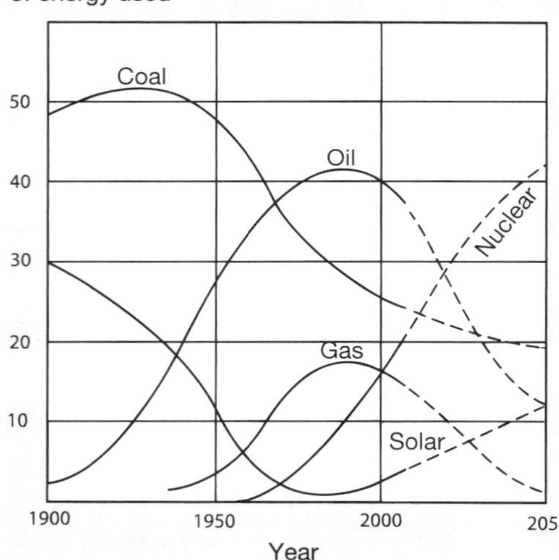

The past, present and projected use of different types of energy in the world.

- **a** In 1900 what percentage of the world's energy came from coal?
- **b** In what year was the percentage of world energy used the same for oil and coal?
- **c** Which two forms of energy are expected to become more important in the world until the year 2050?
- **d** What form of energy is expected to be least important in the world by the year 2050?
- **e** Which two forms of energy follow the same pattern in the way that they have increased and then decreased in importance from 1950 until the present time?

5 Look at the graph shown below and list the major types of energy being used in Papua New Guinea today. Give an example of how each type of energy is used.

M A

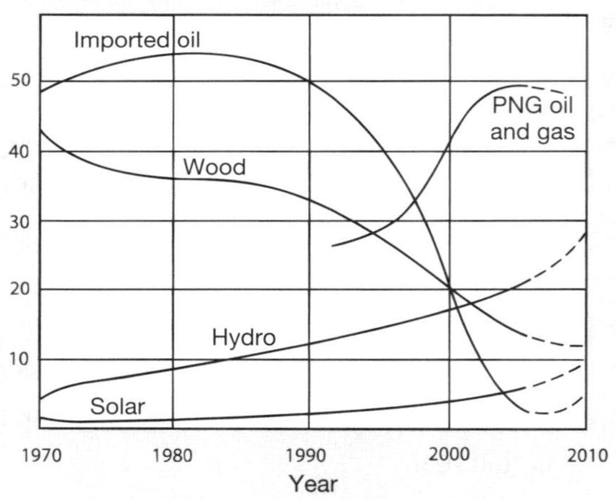

The past, present and projected use of different types of energy in Papua New Guinea.

For you to try

Project: Making a water wheel

1 Collect materials such as aluminium cans, tin cans, corks, plastic bottles, plastic cups, the lids from glass jars, pieces of wire, pieces of string, old bicycle inner tubes, plastic tube or hose pipe, etc.

2 Have a competition in the class to see which group can design and make a model water wheel that will turn under a stream of water. You can make your own stream of water using a siphon made from bucket of water and a piece of plastic pipe or rubber hose.

3 You will need to think carefully about how the wheel is going to work. For example:
   - a What is the best shape for the cups or blades to be turned by the water?
   - b How will the cups or blades be fixed to the axle?
   - c How will the wheel turn on the axle?
   - d How will the axle be supported?

4 The water wheel should be able to stand by itself and not need to be held in your hands. This is so that each wheel can be tested fairly and compared with the others.

5 You may have to try different ideas to see if they work. This is sometimes called **trial and error** and is a good way to find out how things work and solve problems.

6 You will need to agree how much time you have to design and make the wheel.

7 You could have a prize for the water wheel that works the best and invite someone to come and judge the best wheel or students could vote to choose the best.

8 Discuss with other groups why some model water wheels are better than others.

Understanding forces

Many of the activities that we do every day involve lifting, pushing and pulling things. Whenever we are pushing or pulling, stretching or squashing, bending, twisting or tearing we are exerting a **force**.

Scientists say that a force can be a **push** or a **pull**. For example, when we want to move a heavy log of wood or a drum of fuel we push it to make it roll. When we come back from fishing we pull the canoe out of the water.

A force is needed to make something **move**. The heavy log is rolled along the ground and the canoe is moved up the beach. When a jet aeroplane takes off gases are forced out behind and the aeroplane moves forwards. If no force is used then the object will not move. After pulling a canoe up the beach it will not move again until another force is applied.

There are also forces that make things **slow down** or **stop**, like the force of the brakes on a car or the force of a parachute on the air.

Forces can act in any **direction**. When describing forces in science it is often useful to show the direction of the force with an arrow. When something is already moving a force can be used to make it **change direction**.

Forces can change the **shape** of things, like the force of a foot on a soccer ball at the moment it is kicked or the force of a car running into a

Pushing a log along the ground

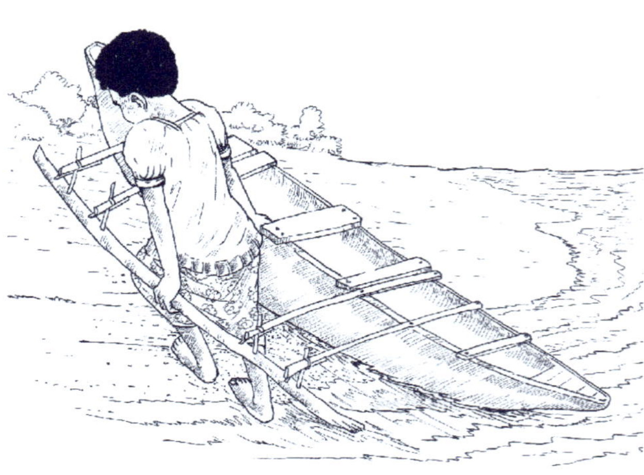

Pulling a canoe out of the water

wooden post. The soccer ball will quickly and easily return to its original shape but this is more difficult with the car.

Some forces act on objects from a distance. The force does not have to be touching the object in order to make the object change. Other forces need to be in contact with the object before they can change the way an object is moving or change its shape.

The table below shows some forces that act at a distance and others that act when they make contact.

The force of the gases coming from a jet engine makes the aeroplane move forwards.

Forces can change the shape of a car in an accident.

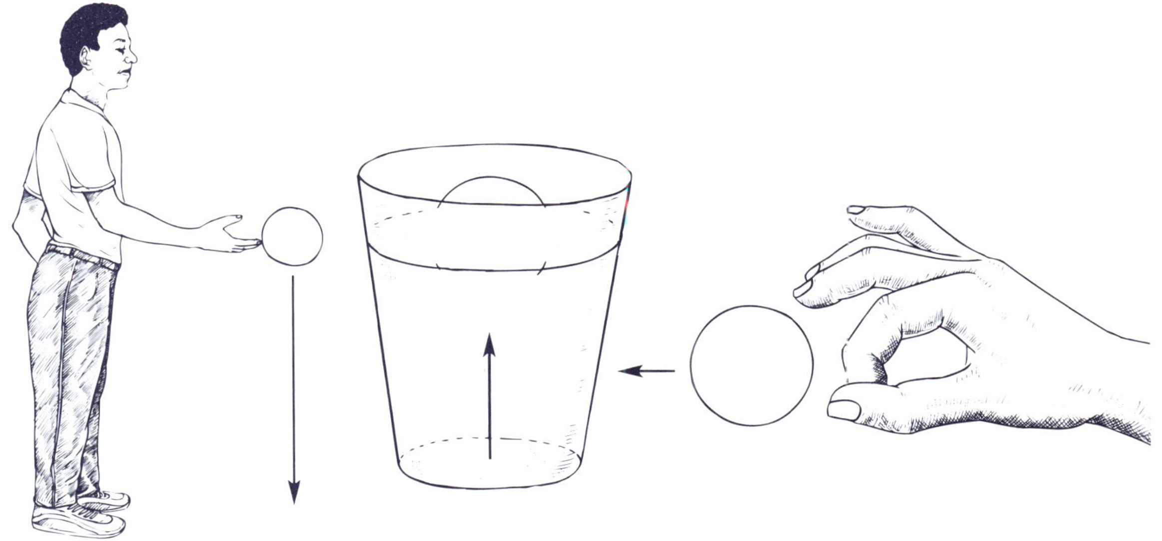

Forces can pull downwards

Forces can push upwards

Forces can push sideways

Different ways that forces act

Forces that act over a distance	Forces that act on contact
Gravitational forces	Explosive forces e.g. a bomb exploding
Magnetic forces	Frictional forces
Electrical forces	Collision forces e.g. a car crash

A force does not always make something move or change its shape. For example, if you stand and push against a big tree, the tree will not move. In this case you are exerting a force on the tree but you cannot see anything happening as a result.

For you to try

1 Investigation: Pushes and pulls

a Collect a piece of rope, bush vine or several laplaps tied together.

b In pairs, stand facing each other. Put the palms of your hands together and push against each other.

c Draw a diagram to explain what happens. Use arrows to show the direction of the forces. Use bigger arrows for bigger forces and arrows of the same size when the forces are balanced.

d In pairs, stand facing each other. Hold each other's hands firmly, lean backwards and pull away from each other.

e Draw a diagram to explain what happens. Use arrows to show the direction and size of the forces.

f Make two equal groups of students. Use the rope or laplaps tied together to have a tug-of-war.

g Repeat with different students in each team or group.

h Draw a diagram to explain what happens. Use arrows to show the direction and size of the forces.

i Try mixing different combinations of big people and small people, boys and girls on each side. What happens? Does it help if people are wearing shoes?

j If you were choosing a group of people for a tug-of-war who would you choose and why?

2 Copy and complete the table below to describe the forces that make things move. One example has been completed for you.

Making things move

Object	How can I make it move?
Ball	kick it, throw it, bounce it
Canoe	
Pencil	
Bush knife	
Desk	
Very light feather	
200-litre drum of kerosene	

3 Copy and complete the table below to describe the forces that make things stop.

Making things stop

Object	How can I make it stop?
Bicycle	
A netball in the air	
A football rolling along the ground	
A person falling over	
Water running in a creek	
Wind blowing into a house	

Types of forces

There are four main types of forces.

The force of gravity

The force of gravity is caused by the earth. Babies soon learn that if they drop something it always falls to the ground. If a pencil rolls off the desk it is pulled downwards and falls to the floor. If you do not harvest young green coconuts for drinking then they will become dry and brown and will fall to the ground. If a bird stopped flying and folded its wings it would also fall to the ground. The force pulling objects down is called the force of **gravity**.

The force of gravity pulling on an object has a special name. It is called the **weight force** or simply the **weight**.

Weight is the pull of gravity that

- holds things on the surface of the earth
- makes things fall back to the ground when they are thrown in the air
- holds satellites and the moon in orbit around the earth.

Magnetic forces

Magnetic forces are caused by magnets. Magnets are usually made of iron and attract things that are made of iron or steel. Magnets are often found inside the plastic strip on the door of a refrigerator or freezer and used to make the door close. Magnets are also used in compasses, electric motors and generators, radios and computers.

Electrical forces

Electrical forces are caused by charged objects. For example, clothes that are made from synthetic or man-made material will try to stick to our bodies. Sometimes the material makes a little crackling noise when we take them off and at night time we might even see a little flash of light.

Elastic forces

Elastic forces can make objects change shape. A piece of foam rubber, a spring and a rubber ball are examples in which elastic forces can act.

Force and movement

Sometimes more than one force acts on an object. For example, when you sit on a chair your weight acts downwards and there is another force, called the **reaction force**, which acts upwards. When you sit still on a chair you are not moving because the two forces are **balanced**. They are equal but act in the opposite direction. Most people do not think about this unless the chair breaks. When a chair breaks your weight will be greater than the reaction force and you will fall to the ground.

When a truck moves along a straight road at a steady speed the forces acting on it are also balanced. There is the force of the engine driving it forward, and the force of friction trying to slow it down.

When the driver puts the brakes on, the truck slows down. This is because the force of friction and the force applied by the brakes are greater than the force from the engine. In this case the forces are **unbalanced**.

Unbalanced forces make things go faster or slower.

Friction: making things stop

Friction is the force that tries to slow things down or make them stop moving. The size of the force of friction depends on the roughness of the two surfaces that are rubbing against each other. Rough surfaces have a bigger area of contact that stops slipping. Smooth surfaces have less points

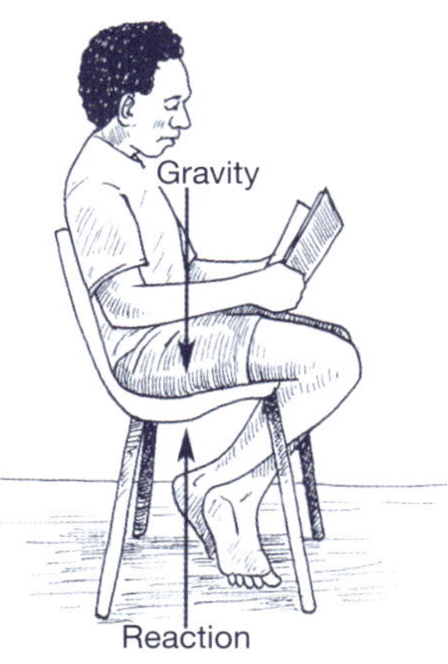

a When you sit on a chair the forces are balanced and you do not move.

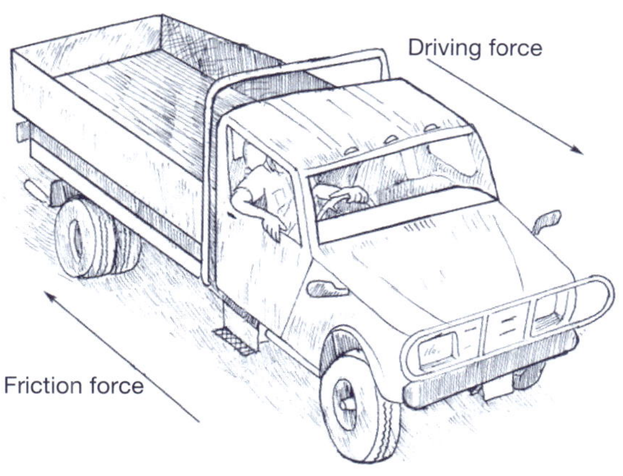

b The truck moves at constant speed because the driving force and the friction force are balanced.

Car tyres and sports shoes have a rough surface to help get a good grip.

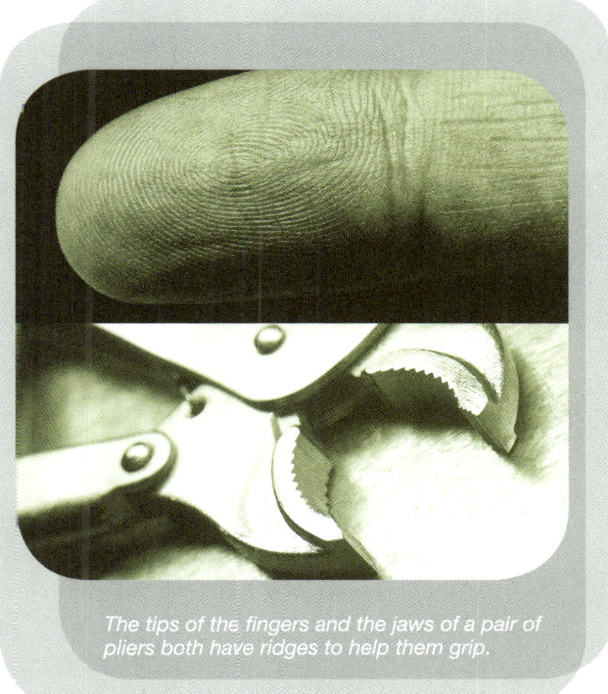

The tips of the fingers and the jaws of a pair of pliers both have ridges to help them grip.

of contact. So rough surfaces have more friction than smooth surfaces. This is why car tyres and sports shoes have a rough tread and must be replaced when they wear out and lose their grip. The tiny ridges on the tips of your fingers that make your fingerprint also help you to grip things and so do the jaws of a pair of pliers.

Friction can also produce heat. If you rub your hands together you can feel the heat produced by friction. The moving parts of a machine can also get hot which can make them stop working. Grease and oil reduce friction so that moving parts do not overheat and stop working.

So friction has advantages and disadvantages. The main advantage of friction is that it allows things to move. For example, when we ride a bicycle friction is needed between your feet and the pedals and between the tyre and the road to make the bicycle move. Friction is needed between the rubber brake blocks and the rim of the wheel to make the bicycle stop. However, friction in the chain and wheel bearings slows you down, which is a disadvantage. Wind resistance also slows you down.

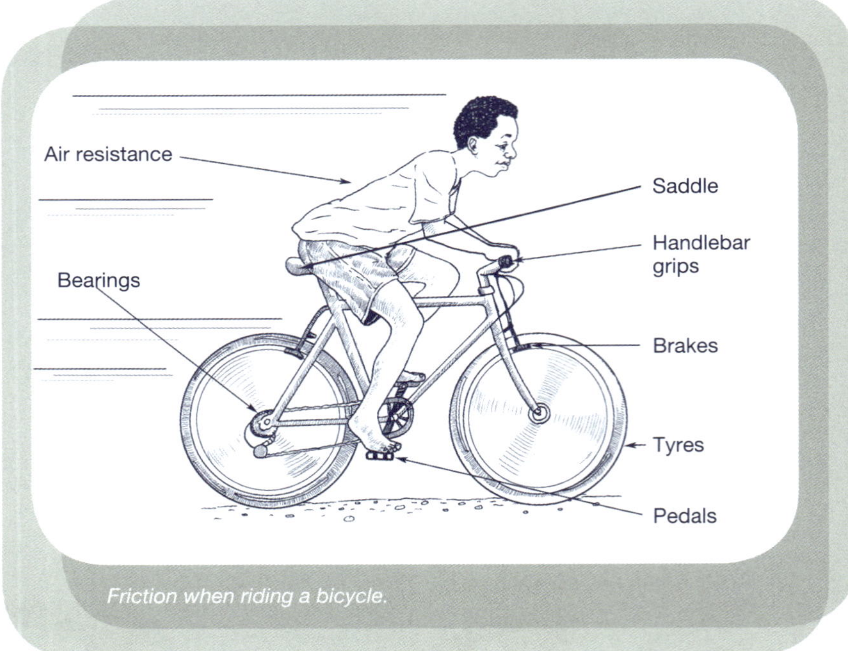

Friction when riding a bicycle.

For you to try

1 Investigation: Investigating forces SR

a Collect a strong cardboard carton, a piece of rope or string, a brick or a rock about the same size, two small rubber balls, a bucket of water, two bar magnets (or one magnet and a nail), one Kina coin.

b Do as many of the following experiments as you can and observe what happens in each case.

c Copy the table on the next page and record your results. Remember to say where the force came from in each case and if the object changed direction.

- **i** Put the rock in the carton and push it across the floor.
- **ii** Put the rock in the carton and pull it across the floor with the rope.
- **iii** Lift the rock from the floor onto a table and back to the floor
- **iv** Drop a rubber ball onto the floor.
- **v** Place a rubber ball in a bucket of water.
- **vi** Push a small rubber ball to the bottom of a bucket of water and let it go.
- **vii** Cut a piece of cardboard the same size as a one Kina coin. Hold both the coin and the piece of cardboard at the same height and let them go at the same time.
- **viii** Roll two balls along the floor from different positions so that one hits the other.
- **ix** Bring the magnet close to the nail. If you have two magnets bring different ends of the magnets together.

d Copy the table on page 109 and record your results. Remember to say where the force came from in each case and if the object changed direction.

Object and force applied	Where did the force come from?	Did the object change direction?

2 Investigation: Friction in a pair of thongs

a Collect several different pairs of thongs, some mud (or find a muddy place close by), sand, a bar of soap and water.

b Carefully observe the bottom of the thong that touches the ground and the top that touches the sole of your foot. Draw labelled diagrams to show the pattern on both sides. What difference do you notice between the top and the bottom of the thongs and between different kinds of thongs?

c Describe what it is like to walk in thongs in the following conditions:

i Walk in clean, dry thongs with clean, dry feet.

ii Walk in the thongs with wet, muddy feet. There should be mud between the sole of your foot and the thong and between your toes.

iii Sprinkle some sand on the muddy thongs and again try walking in them.

iv Rinse the mud off your feet and the thongs with water, then wash your feet with soap and leave them wet and soapy. Again walk in the thongs.

v Finally, rinse and dry your feet when you have finished.

d Describe the results of your investigation and what you have found out.

e Explain why there is a pattern on the top and bottom of the thongs. What happens when the bottom of the thongs becomes worn?

3 What is a force? For each of the pictures in the diagram, describe the main type of force acting and also describe what this force is doing. Remember to use the words pushing and pulling.

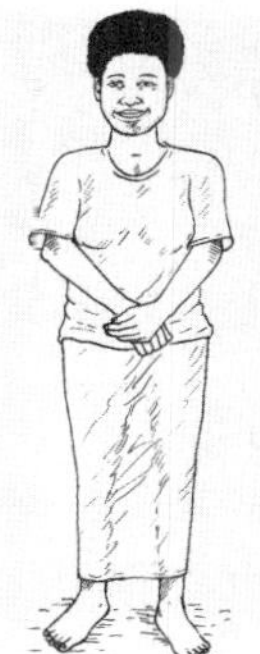

4 Draw a diagram of each of the following to show the forces that are in action. Label the forces involved as a push or a pull.
   - **a** A tug-of-war between two teams
   - **b** Hoisting a flag on a flagpole
   - **c** Pulling up the anchor of a boat
   - **d** A person walking down the road
   - **e** Flying a kite
   - **f** Moving a rock with a crowbar
   - **g** Catching a fish on a hand line
   - **h** Firing an arrow or a spear

5 Explain what would happen to a car or truck that has worn tyres that are almost smooth or bald:
   - **a** when the car tries to stop quickly
   - **b** when it is wet

   In many countries it is against the law to drive a car or truck with very worn tyres. Why is this?

Elastic objects

When you kick or push on a rubber ball it changes shape. When you stop pushing, it returns to its original size and shape. Objects that return to their original size and shape after a force has been removed are described as being **elastic**. A rubber band, a foam rubber mattress and an inner tube from a bicycle or truck are all elastic.

Some objects do not return to their original shape when the force is removed and they are described as being **inelastic** or plastic. For example, soft clay and plasticine are inelastic.

The force of a foot changes the shape of a soccer ball but the ball quickly returns to its original shape.

For you to try

1 Investigation: Finding out about elasticity

Collect the following: an eraser, a rubber band, a lump of clay, a blown-up balloon, a plastic syringe, or a pump for bicycle tyres or sports balls.

a Compress or squeeze the eraser and then the clay with your fingers. What happens after you remove the force, or stop squeezing? Next, stretch the eraser and the clay by pulling on each. What happens to each when you stop stretching?

b Stretch a rubber band a number of times. Does it feel elastic? How do you know? What happens if you overstretch it?

c Pull back the piston of a syringe or the handle of a pump. Put your finger over the end and then push in the plunger or the handle. Can the air be compressed? What happens when you release the plunger or the handle? Does the air have some elasticity? How do you know?

d Compress or squeeze a blown-up balloon. Can the balloon be compressed? What happens when you stop squeezing the balloon? Does the balloon have elasticity? How do you know? What happens if you squeeze too hard?

Draw a table to briefly describe the tests used on different objects, and the results.

2 Copy and complete the following table of objects that are found around your home or school. Classify the objects according to whether they are elastic or inelastic. **A**

Object	Elastic	Inelastic

Simple machines

A **simple machine** is something that helps to make work easier. For example, it is easier to move a bag of cement or a load of rocks if you put them in a wheelbarrow than if you try to carry them in your arms. So we say that the wheelbarrow is a simple machine.

The main types of simple machines are:

- levers
- the wheel and axle
- inclined planes
- pulleys and gear wheels.

Using a wheelbarrow.

The spade is a lever.

The steering wheel of a car, truck or tractor consists of a wheel and axle.

These machines can do three things to help us. They can:

1 move a heavy load with a small force
2 make things go faster
3 change the direction of a force.

When we use the word 'machine' we might also think about things like a sewing machine, a bicycle, a tractor or the scales used to weigh food. These are machines but they are more complicated. Complicated machines are often made up of a number of simple machines.

Levers

Cars and trucks have a gear lever in order to change gear. When we get a puncture in a car or a truck tyre, we need a tyre lever to remove the tyre from the wheel rim. Other tools which work in the same way are a crowbar, a pinch bar, a spade, a wheelbarrow and a bottle opener. All of these tools are simple machines called **levers.** A lever is usually

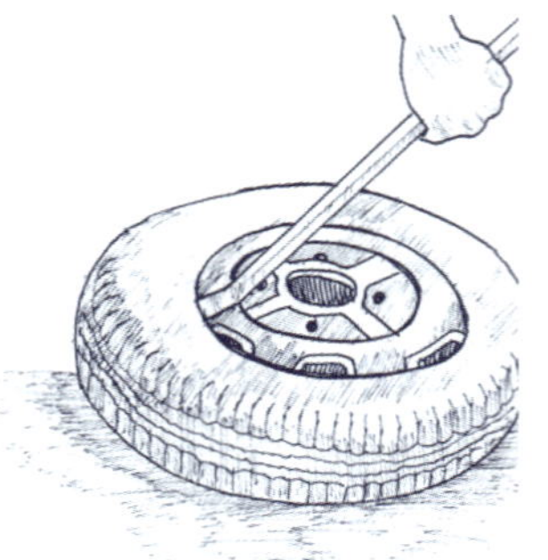
a Removing a tyre from a wheel rim with a tyre lever

b Moving a log with a crowbar

c Moving a rock with a crowbar

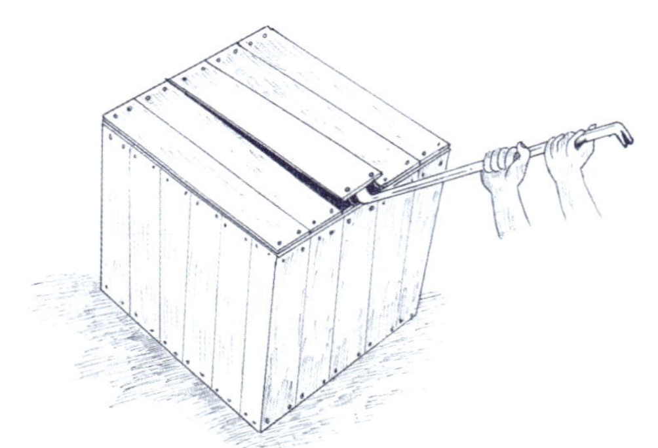
d Opening a packing case with a pinch bar

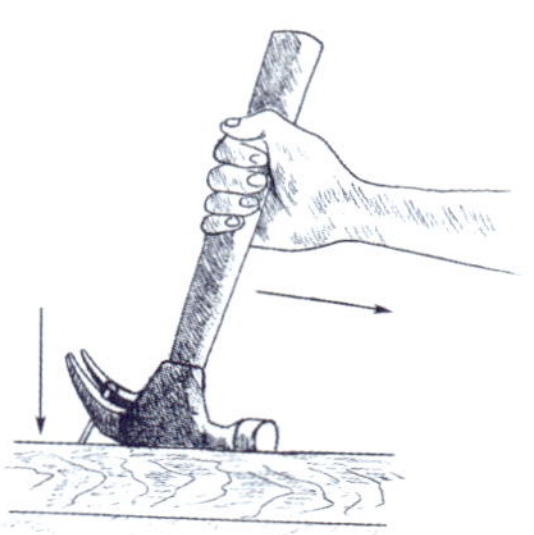
e Removing a nail with a claw hammer

f Opening a bottle with a bottle opener

a strong, stiff bar that can turn or **pivot** around a fixed point called a **fulcrum**. It is easy to see why a crowbar and a bottle opener are called levers.

A wheelbarrow lets us lift a heavy load easily because only a small force is required. The diagram shows the large weight force of the load and the small lifting force of the man. The pivot point in the wheelbarrow is the axle.

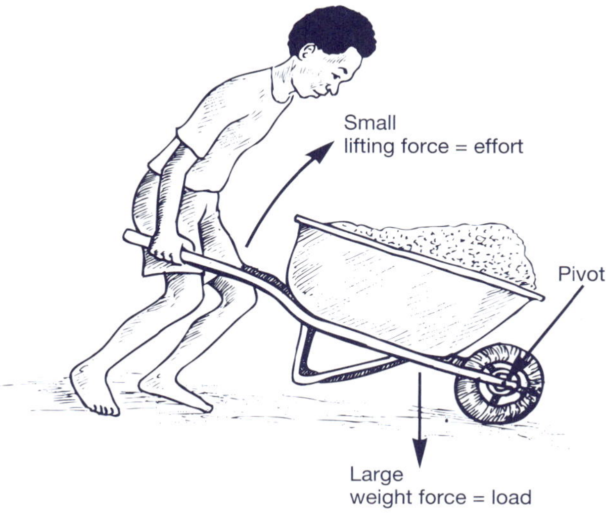

The wheelbarrow is a simple lever in which there is a force advantage.

The easiest way to use a wheelbarrow is to put the load close to the pivot point and to have your lifting force, or effort, far from the pivot.

Some simple machines are made of double levers. The bamboo tongs that are used for lifting hot mumu stones or turning the food cooking in a fire are an example of a double lever. Other examples are scissors, pliers, tin snips and forceps or tweezers.

Bamboo tongs are an example of a double lever.

For you to try

1 Investigation: Opening and closing a door

a You will need a door that is fixed with hinges that you can reach easily.
b Try opening or closing the door by pushing it gently near the hinge.
c Try opening or closing the door by pushing it gently near the handle.
d Which is easiest? When do you have to push hardest?

2 Investigation: Making bamboo tongs

a Collect some bamboo, ribs from a coconut frond or similar material and a bush knife.
b Make a pair of bamboo tongs that can be used to lift hot mumu stones or turn the food cooking in a fire. Try using different lengths of bamboo.
c Which one works best? Why?

3 Investigation: Finding out about levers

a Collect an old powdered milk tin, a 20t coin and different kinds of levers like knives, forks and spoons of different sizes and made of different materials, both metal and plastic.

b Try opening the milk tin using only your hands.

c Now try opening the tin with the 20t coin and with each of the levers. Which ones work the best?

d Draw a table to show your results.

---

4 Investigation: Using a crowbar

a You will need a crowbar or a suitable piece of strong timber, a short, thick piece of round timber and something large and heavy to move. You also need two chairs for the last part.

b With your teacher, try moving a log or a rock or some other large, heavy object using a crowbar or a suitable piece of timber.

c Try using a pivot or fulcrum to change the position of the crowbar or piece of timber.

d Take care that you do not trap your hands or feet or hurt other people.

e If you have two chairs you can also try doing the investigation shown in the diagram which works in the same way.

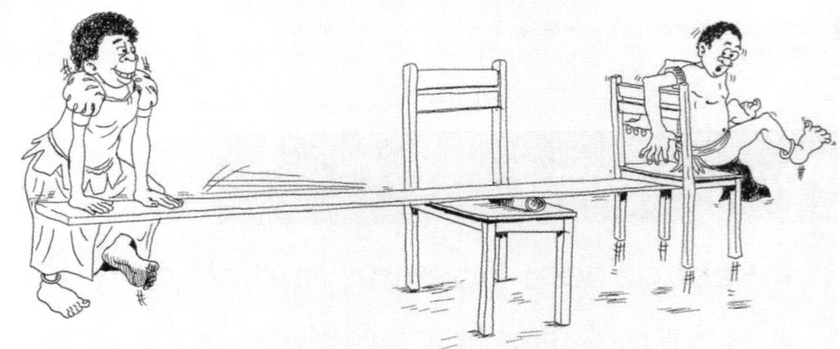

5 What is a see-saw? How does the see-saw work? What happens when people of different weights sit at different positions? If you can, make a see-saw using a long plank of wood and a piece of the trunk of a tree as a pivot or fulcrum and try out your ideas.

6 Draw a diagram of a pair of scissors or a pair of pliers and explain why they are called double levers. Label your diagram.

Wheel and axle

A simple machine that uses the principle of the lever is the **wheel and axle**. The screwdriver and the steering wheel of a car or truck are both examples of the wheel and axle. The wheel and axle is made of two wheels which are joined together. The smaller wheel is called the **axle**.

A small force is used to turn the larger wheel. This causes a large force to act on the smaller wheel or axle. In this way a small force applied to the wheel can move a large load attached to the axle. This makes it easier to steer a car or truck and to tighten a screw using a screwdriver.

Inclined planes

It is much easier to roll a 200 litre drum of fuel up a ramp on to a truck than to lift it. One person can roll a drum up a ramp but it is impossible for one person to lift a drum of fuel on to a truck. A sloping plank or ramp like the one in the diagram is called an **inclined plane**. An incline is a slope and a plane is a surface.

To move something up a gentle slope or incline requires less force than to move something up a steep slope or to lift it vertically. Steps, stairs and ladders are a type of inclined plane and so are winding roads that go up mountains. The road to Sogeri from Port Moresby and many roads in the highlands are examples of winding zigzag roads that make it easier to go up and down.

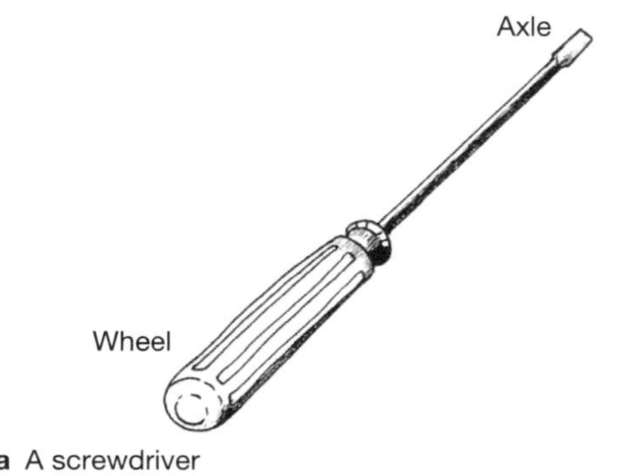

a A screwdriver

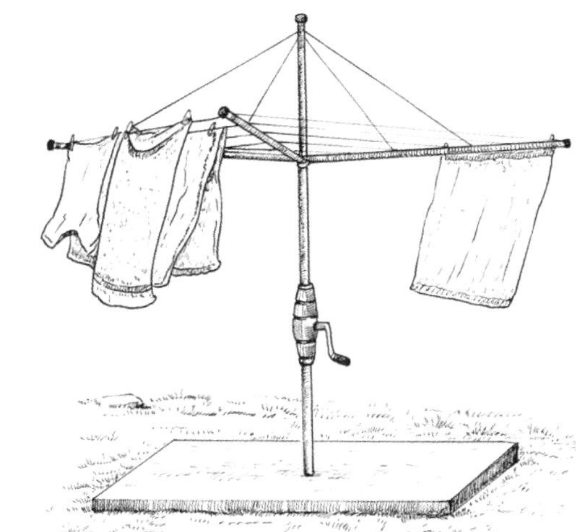

b The winder on a rotary clothes hoist

c A door knob

Simple wheels and axles.

Rolling a drum up a ramp.

The road from Port Moresby to Sogeri is a winding zigzag road.

In order that they are not too steep, these roads must keep changing direction across the hillside and so there are often sharp, hair-pin bends. This kind of winding mountain road is also called a **switchback** road.

Screws, nuts and bolts are also examples of an inclined plane. A screw has a **thread**, which is an inclined plane that turns round and round in a spiral. A small force applied to the head of the screw can produce a large force to hold things together. Electrical appliances like refrigerators, freezers and washing machines often have screws on the legs at the front. These screws can be adjusted to make the appliance stand level on the floor. The jack that is used to change the wheel of a car usually has a screw thread to lift the weight of the car.

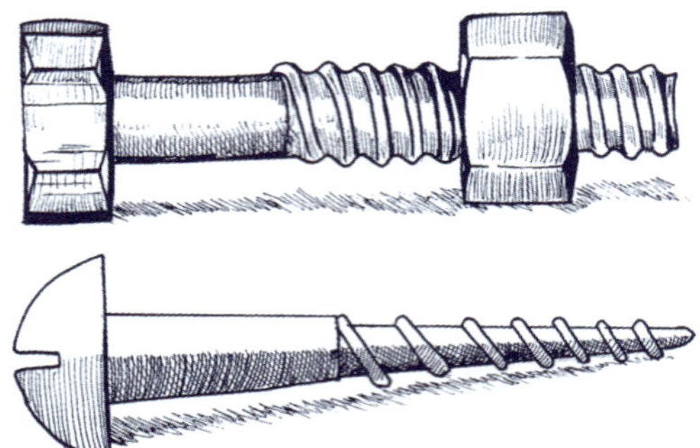

Screws, nuts and bolts have a thread which is an inclined plane.

A **wedge** is a simple machine made up of two inclined planes placed back to back. It is used to push things apart. For example, the head of an axe can be used as a wedge to split a log.

An axe can be used as a wedge to split a log.

For you to try

1 Investigation: Using an inclined plane

a Collect a 20 litre drum of kerosene and a plank. Make sure that the lid is on tight so that the kerosene does not come out.

b Try rolling the 20 litre drum of kerosene up a ramp made with the plank of wood.

c Explain why it is easier to do this than to lift it.

2 Investigation: Walking up hills

a If you live near a steep hill try walking straight up the slope and then try making a zigzag path across the slope.

b Which way is the shortest? Which way is the easiest?

c Explain why it is easier to walk up a hill by a winding zigzag path rather than walking straight up.

3 What is an inclined plane? Give some examples.

Pulleys

Pulleys are simple machines that can change the direction of a force. Pulleys usually consist of one or more wheels and a length of rope, although sometimes they use a chain or a belt. A **single fixed pulley** allows you to lift an object upwards by pulling down on a rope. You can easily make a bucket shower using a single fixed pulley. You can lift an object which weighs as much as you by using a single pulley. The branch of a tree can also be used as a single fixed pulley. The branch acts like a pulley wheel although there will more friction than using a real pulley wheel.

A single **movable pulley** can be used to give a force advantage. You need less force to move the load with a pulley, but you need more rope. A single movable pulley does not change the direction of a force. The load will move in the same direction as you pull but you will use a smaller force.

A single fixed pulley can be used to make a bucket shower.

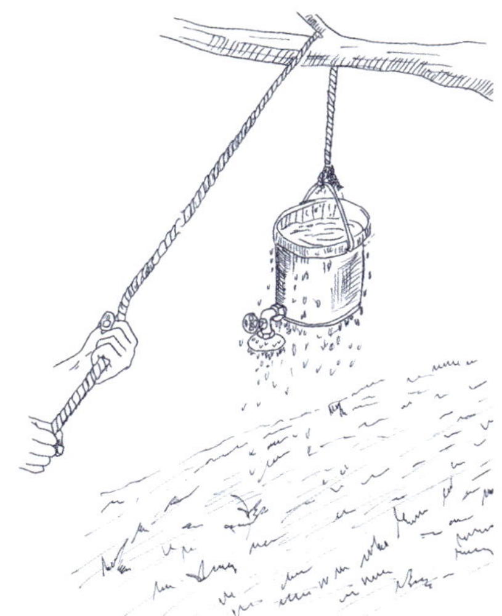

The branch of a tree can be used as a simple fixed pulley to make a bucket shower.

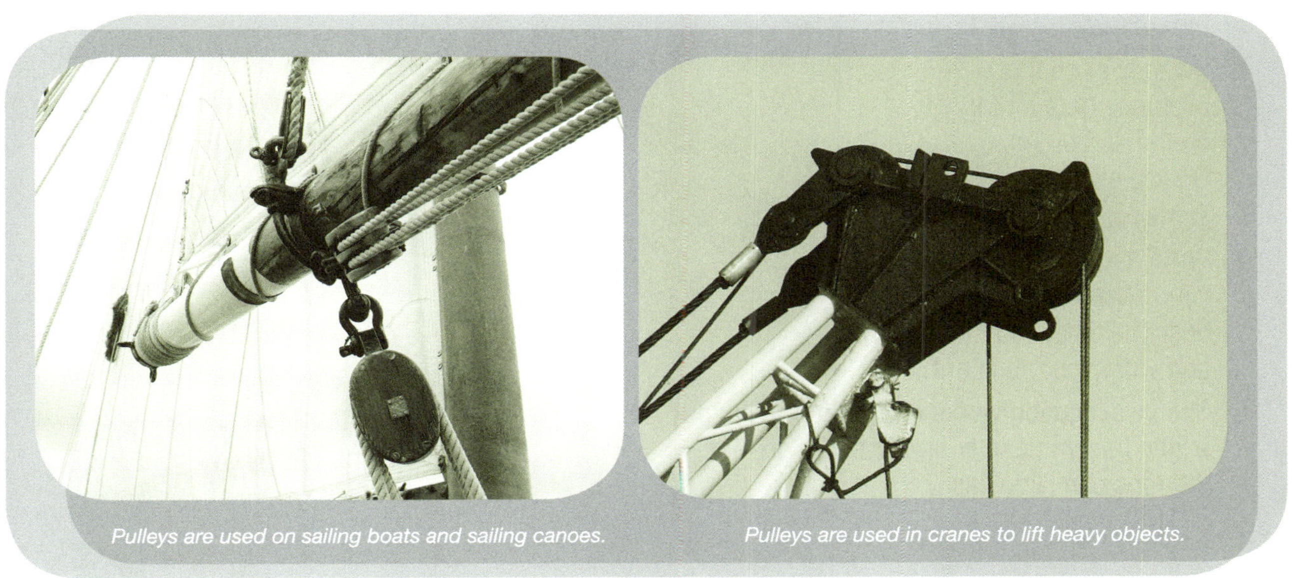

Pulleys are used on sailing boats and sailing canoes.

Pulleys are used in cranes to lift heavy objects.

Pulley blocks with more than one pulley wheel can be used to lift a heavy load, but you need a lot of rope to lift the object only a little way. Pulley blocks are used to control the sails on sailing boats and are used on cranes to lift heavy objects.

For you to try

Investigation: Four against one SR

1. Collect two strong, smooth sticks or broom handles and a long rope.
2. Work in groups of five.
3. Tie one end of the rope to one of the sticks and then loop it from one stick to the other in a zigzag pattern.
4. Two people hold each stick and try to pull them apart while you pull on the end of the rope.
5. What happens? Can one person pull against four others?
6. Swap over so that everyone gets a turn on the end of the rope. Do you always get the same result?
7. Write a report of your investigation to describe what you did and what you found out.
8. The investigation works even better if you put a little talcum powder on the sticks so that the rope can slide easily.

Pulleys and gear wheels

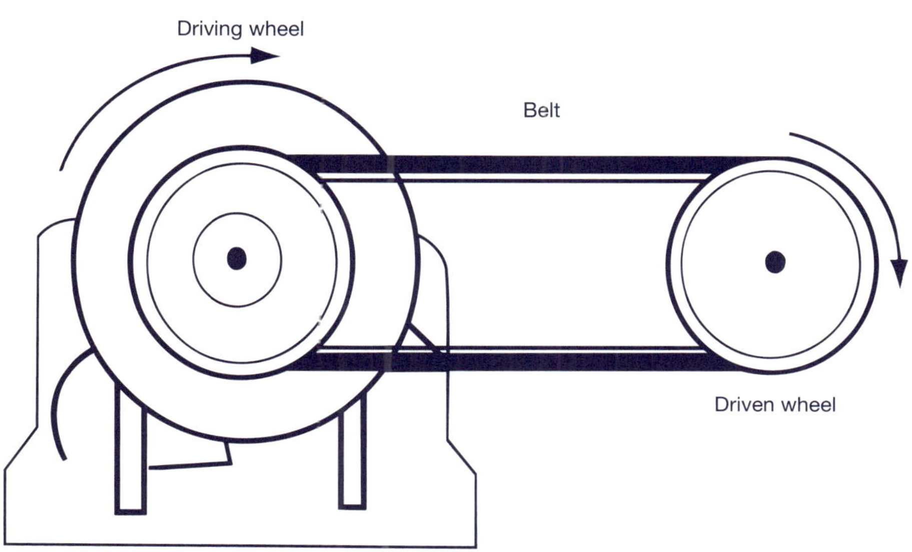

Levers, the wheel and axle, the inclined plane and pulleys are simple machines that move things by sliding or lifting. However, some machines move objects by turning wheels and cogs. For example, egg beaters, sewing machines, bicycles, concrete mixers, engines and lathes all depend on **gears** which interlock or

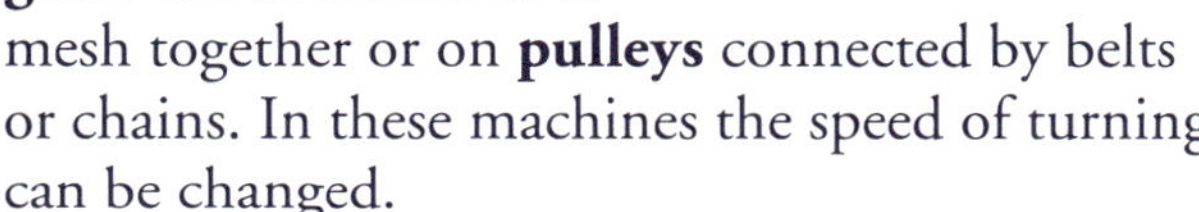

mesh together or on **pulleys** connected by belts or chains. In these machines the speed of turning can be changed.

A belt and pulley system has a driving wheel, a driven wheel and a connecting belt or chain. The driving wheel is attached to the driving force and gives the effort. The driven wheel is attached to the load. For example, in many trucks and older cars the bottom pulley is driven by the engine. The fan belt goes around the bottom pulley and drives the fan and the water pump. The fan pulls air through the radiator and the water pump forces water round the engine which both help to cool the engine. The fan belt also drives the alternator which charges the battery. Some modern cars have an electric fan for the radiator that works separately and is not driven by the fan belt.

A bicycle has a **chain** and toothed **gear wheels**. One of the advantages of a chain is that it is less likely to slip than a belt, but the chain can still come off the gear wheels. In a bicycle the back wheel must turn quickly when you push on the pedals. So the gear wheel to which the pedals are attached is bigger than the gear or sprocket on the back wheel.

Because they transfer energy from one wheel to another, belts and chains can slip or break. To overcome this problem, engineers have made special gear wheels with matching teeth around the edges. The teeth on one wheel fit or **mesh** into the teeth of another wheel. This idea is used in the gearbox of cars and trucks.

Belt and pulleys on a car engine

As with pulleys, if the driving wheel is bigger, then the wheel that is driven moves with increased speed. On the other hand, a small driving wheel turning a large wheel gives a force advantage. In both cases the driven wheel will turn in the opposite direction to the driving wheel.

Meshing cogs or gears prevent the slipping that can occur in a belt system.

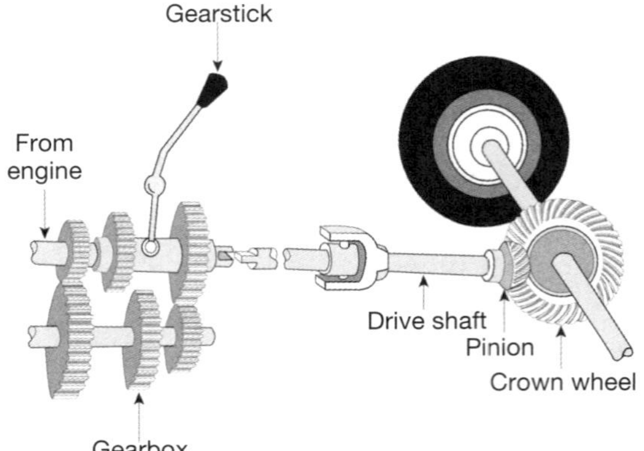

The gearbox and differential on cars and trucks.

For you to try

1 Investigation: How do gears work? SR

a Collect two plastic lids of different sizes, a rubber band, two nails and a small piece of wood. You will also need to use a hammer.

b Using a hammer and nail, carefully make a hole exactly in the centre of each lid.

c Move the nail around in the hole to make the hole a little bit bigger. You want the lid and nail to be like a wheel and axle so that it will turn easily, but don't make the hole too big.

d Fix each lid to a piece of wood using a nail. Just tap the nail a little way into the wood so that you can easily remove the nail if you want to.

e Stretch a rubber band around both lids so that it is tight.

f Place a mark on the top outer edge of each lid.

g Turn each lid at a time and observe what happens to the other lid.

h When you turn the big lid once, how many times does the small lid turn?

i Write a report of your investigation to say what you did and what you found out.

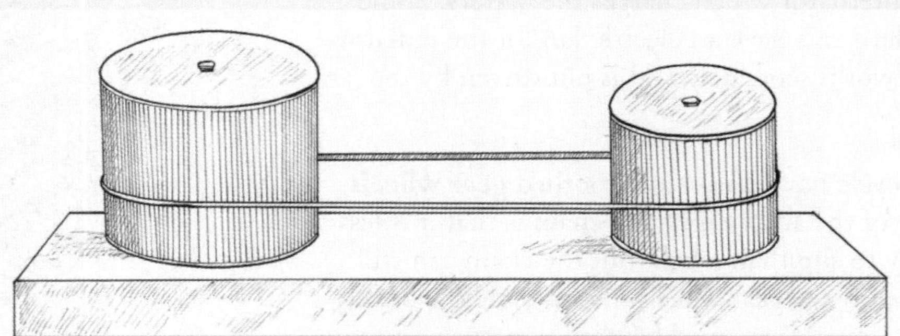

2 What is a simple machine? List four types of simple machines. Give an example of each type of machine.
3 Make a poster to show a simple machine. Label the poster to show how the simple machine works.
4 Make a model of a simple machine to show how it works.
5 Think about some work that needs doing in your home or community. Draw a labelled diagram or poster to explain how this work could be carried out more easily using a simple machine. The following examples may help you:

Example 1: When removing a nail from a piece of wood using a claw hammer, it often helps to put a small block of wood under the hammer to make the fulcrum higher. This gives more leverage as the nail comes out and helps to pull the nail out straight. If the nail is long and very hard to pull out you may need to change the size of the block of wood more than once.

Example 2: When undoing the nuts of a wheel to change a tyre you can slide a piece of galvanised water pipe over the handle of the wheel brace in order to get more leverage.

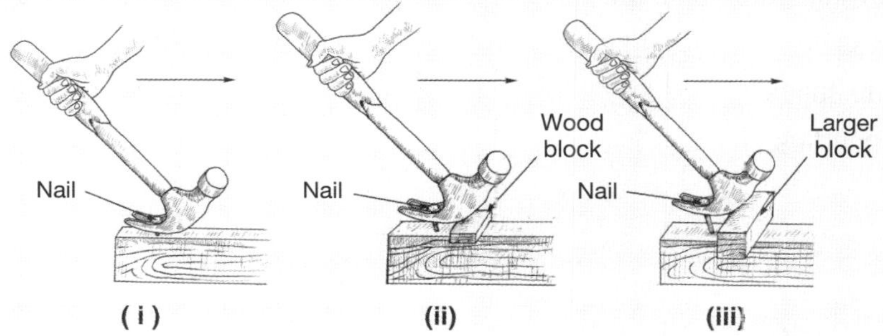

Removing nails using a claw hammer by using a small block of hardwood under the hammer as a fulcrum.

Using a piece of water pipe to help change a wheel by making the handle of the wheel brace longer.

Summary questions

1 Which of the following best describes solids, liquids and gases?

A the nature of matter

B the properties of matter

C the states of matter

2 Which of the following statements are true?

i Matter has mass

ii Matter takes up space

iii Matter has properties

A i only

B ii only

C i and ii only

D i, ii and iii

3 Which of the following best describes the different things that we use in our daily lives, from the largest number to the smallest number?

A solids, liquids, gases

B liquids, solids, gases

C solids, gases, liquids

D gases, liquids, solids

4 Which of the following best describes the volume and shape of matter?

	Fixed volume	Fixed shape
A	solid, liquid	gas
B	solid, liquid	solid
C	solid, liquid	liquid
D	liquid, gas	solid

5 Which of the following best describes the four changes of state?

	Melting	Evaporation	Condensation	Solidification
A	solid → liquid	liquid → gas	gas → liquid	liquid → solid
B	solid → liquid	gas → liquid	liquid → gas	liquid → solid
C	solid → liquid	liquid → gas	liquid → solid	gas → liquid
D	solid → gas	liquid → gas	gas → liquid	solid → liquid

6 Which of the following best describes something that contains two separate substances?

A solution

B solute

C solvent

D mixture

7 Which of the following best describes the way to separate an insoluble solid from a liquid?

A decanting

B decanting or filtering

C dissolving

D dissolving and evaporating

8 Which of the following best describes the way to separate a soluble solid that is dissolved in a liquid?

A decanting

B dissolving

C evaporating

D filtering

9 Which of the following changes is most likely to occur?

A insoluble substance → solution

B solute → solvent

C solution → sediment

D suspension → sediment

10 Which of the following best describes different kinds of energy?

i Energy can make things move

ii Energy can make things hot

iii Energy can make sound

A i and ii only

B ii and iii only

C i and iii only

D i, ii and iii

11 Which of the following are renewable sources of energy?

A firewood and hydro-electricity

B coal, oil and gas

C wind and fossil fuels

D batteries

12 Which of the following statements about forces are true and which are false?

a A force can be a push or a pull.

b Something can move if no force is applied.

c Gravity is a force.

d A force can act in any direction.

e Every force makes something change direction.

f An elastic force can make something change shape.

g Magnetic forces are usually caused by magnets.

h Electrical forces can make things move.

i Friction is a force that is always a disadvantage.

13 Which of the following best describes the main way in which a wheelbarrow helps us as a simple machine?

A moves a heavy load with a small force

B makes the load go faster

C makes the load lighter

D changes the direction of a force

14 Which of the following best describes the main way in which the gears of a bicycle help us as a simple machine?

A moves a heavy load with a small force

B makes us go faster with less effort

C makes us go slower

D changes the direction of a force

15 The passage below is a summary of the main ideas of this chapter. Copy and complete the passage in your book. (L)
Using the words in the list, find the words that are missing. You can use each word only once.

classified	decanting	dissolves
energy	force	form
friction	levers	machines
move	pulls	substances
solution	suspension	work

All substances can be ____________ into groups known as solids, liquids and gases depending on their properties. Some ___________ can mix together and some cannot mix together. A mixture is formed when one substance ___________ in a liquid to form a solution. A substance that is dissolved in a _________ can be separated from the liquid. A substance that will not dissolve in a liquid may form a ____________. Some mixtures of substances that do not dissolve can be separated by __________.

Energy is the ability to do ________. Energy can be changed from one _____ to another. Different sources of _______ are used in the home. Simple _________ can be used to do work in the home and the community. There are five different types of simple machine: ______, pulleys, axles, inclined planes and gears.

A _______ can be a push or a pull. Forces make objects ______. Friction is needed to allow things to move, but ________ will also make things slow down or stop. Gravity is the force that _______ an object towards the earth.

For you to try

Project: Making a bucket shower (PD)

1 Collect a bucket, a length of hose or plastic pipe, several empty soft drink cans, some string, a length of rope, a hammer and a nail. You will also need a tree with a branch that you can reach that is strong enough to carry the weight of the bucket of water, or some other support.

2 Use the hammer and nail to make twelve holes in the rim of the can.

3 Fill the can with water and test it to see that the water runs out easily.

4 Push one end of the hose into the can. Put the other end of the hose into the bucket and tie it to the handle of the bucket so that the end of the hose will stay on the bottom of the bucket.

5 Fill the bucket with water and make sure that the hose and can also fill with water.

6 Using the rope and the branch of a suitable tree as a pulley, lift the bucket and tie the end of the rope carefully so that the bucket stays in the same position. You should be able to reach the bucket.

7 Quickly and carefully lift the can out of the bucket. Make sure that the end of the hose stays in the can.

8 If you hold the can below the bucket the water should start to flow or siphon from the bucket through the holes in the can.

9 Try cans with a different number and position of holes to get the best shower.

10 As an alternative to a soft drink can you could try using a plastic bottle. Cut a hole in the top so that the hose fits tightly. Make holes in the bottom of the bottle to let the water come out.

a Using a hammer and nail to punch holes in the bottom rim of a soft drink can.

b Using the shower suspended from the branch of a tree.

4

Earth and beyond

Chapter summary

In this chapter you will have an opportunity to:

- find out about different stories that explain the formation of the earth
- make a model of the earth's structure
- find out about the way in which rock and soil is formed.
- find out about the patterns of the stars and the phases of the moon.
- find out about the importance of the moon and the stars in the daily lives of people.

Syllabus references

Strand: Earth and beyond

Sub-strand: Our Earth and its origin

Outcomes:

6.4.1 Investigate the earth's structure and describe the formation, composition and cycling of rock.

6.4.2 Identify and describe familiar events such as star patterns and moon phases.

Key facts

- There are different ideas about the way in which the earth was formed.
- The earth has three main layers: the crust, mantle and core.
- Rock can be formed in different ways; for example, from volcanoes and from the sediments that are carried in water and settle on the bottom of the sea.
- The materials that make up rock can move in a cycle. Rock can be slowly worn away by water and wind and the small particles can be carried by water and settle down as sediments. Over thousands of years the sediment can form more rock.
- Soil is formed from small particles of rock and the living things that have died and rotted. Soil can be carried away by water and wind. This process is called soil erosion.
- Stars form a pattern in the sky that changes regularly with the seasons.
- The moon has different phases that change in a cycle every 28 days.
- People use the stars and the moon to find their way or navigate and to measure time. For example, when setting the time for important events, when planting and harvesting and when planning trading expeditions.

Our Earth and its origin (PD) (SS)

The formation of the earth

In Papua New Guinea today there are many differences between the provinces. All the coastal provinces have islands, but some of these islands have no rivers. The Western Province has very large areas of flat land and the biggest river in the country. The highlands provinces have rugged mountains and fast-flowing rivers, but less flat land. The way in which the land was formed and how these differences came about is very interesting. People in Papua New Guinea often have stories that explain how the earth was formed or how a particular mountain, river, lake or cave was formed. These stories are called **creation stories**.

Scientists also have explanations of how the earth was created. Scientists think that the earth formed from a large cloud of very hot dust and gas around the sun about 4600 million years ago. As the gases and dust cooled, the earth formed. It gradually became smaller to form a ball-shape of molten rock with a solid surface.

The earth has been cooling for millions of years, but the inside is still very hot. In some places the hot rocks are found near the surface. The hot rocks cause steam, hot gases, boiling water and mud. These areas are called **geothermal areas** and are found in many parts of Papua New Guinea, for example, Lihir Island in New Ireland, Lou Island in Manus and Kairiru Island in the East Sepik.

There are many differences between the provinces of Papua New Guinea. (a) the highlands (b) the lowlands

Geysers and hot springs tell us that it is very hot inside the earth.

Boiling mud at Rotorua.

In East and West New Britain, the megapode bird lays its eggs in mounds of soil which are kept warm by heat that comes from hot rocks. The warm soil helps to incubate the eggs so they can hatch. At Salamo in Milne Bay the hot springs are used for bathing, washing clothes and even cooking.

Temperature of the earth's crust

Depth below surface (km)	Temperature (°C)
0	20
1	35
2	50
3	65
4	80
5	95
6	110
7	125

We also know that the earth is hot inside because when miners go down the deepest mines, the temperature rises. For example, at a depth of seven kilometres, the temperature is about 125°C. However, miners cannot work at temperatures above about 45°C. We know that there is still molten rock deep inside the earth because it sometimes comes to the surface when a volcano erupts.

The earth, water and air

The earth is made up of land which consists of rocks and soil, and water. Only about 30 per cent of the earth is land and about 70 per cent is sea water. Because there is more water than land on our planet, the earth is sometimes called the **water planet**. We can see this more clearly if we look at pictures of the earth taken from space. Most of the surface of the earth is covered with water.

Water on the planet earth is one of the reasons why plants and animals are able to live here. Another reason is that the earth is surrounded by a layer of gases called the **atmosphere**. The main gases in the atmosphere are:

- nitrogen (78 per cent)
- oxygen (21 per cent)
- carbon dioxide (less than 1 per cent)

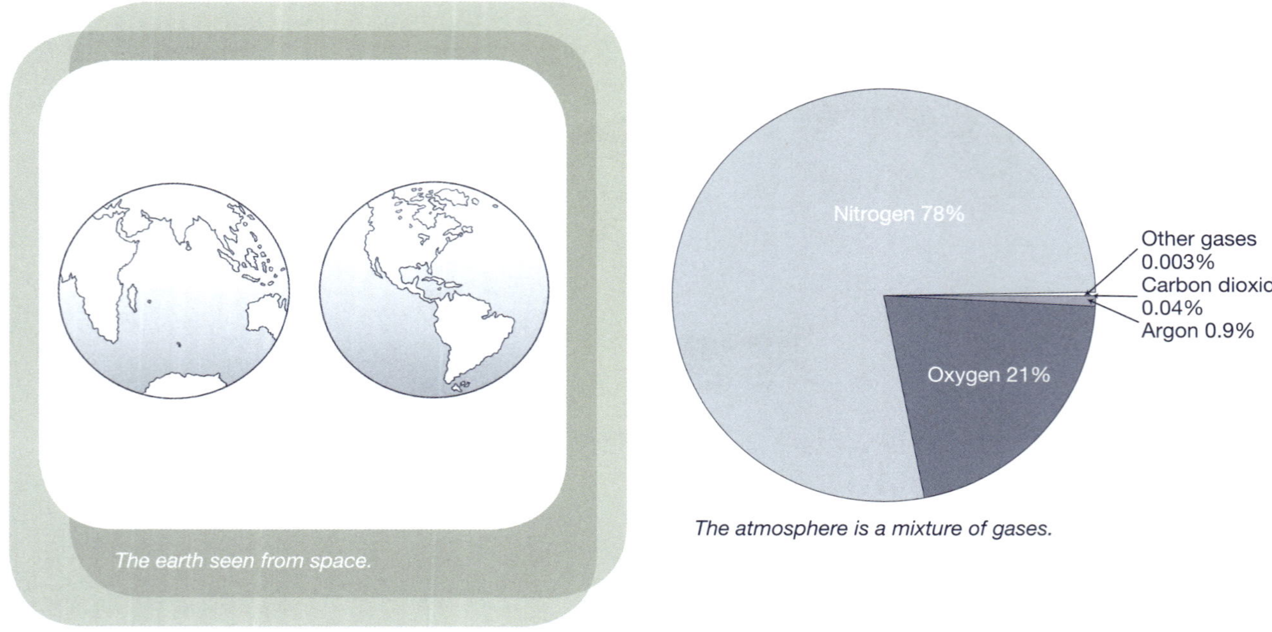

The earth seen from space.

The atmosphere is a mixture of gases.

The atmosphere is very important to living things. In the daytime plants are able to take in the carbon dioxide from the air and use it during the process of **photosynthesis**. This process produces food and also gives out oxygen. During the night-time plants can also give out some carbon dioxide. The oxygen in the air that is breathed in by animals is needed to get energy from food during **respiration**. They then breathe out carbon dioxide which the plants can use. In this way plants and animals depend on each other and are constantly exchanging oxygen and carbon dioxide with the atmosphere.

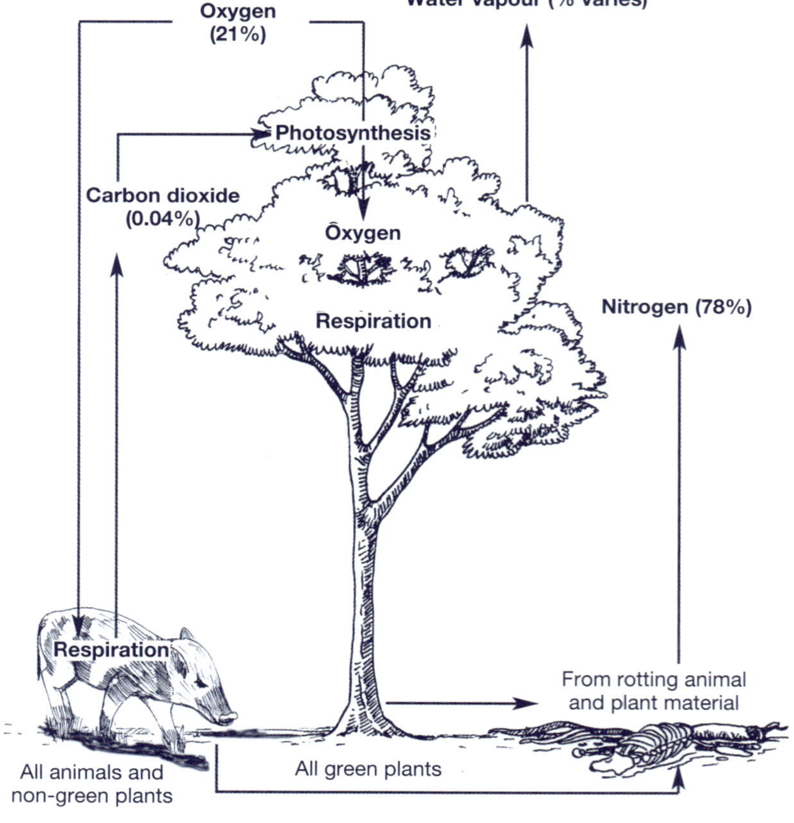

All living things depend on each other for gas exchange.

For you to try

1 Find out about local stories that explain the formation of the earth or the formation of something in the local area, like a mountain, an island, a swamp, a lake or a river. You could do this by inviting a village elder to come to your school or by visiting them and talking to them.

In small groups, make a poster, write a poem or make a play so that you can show others these creation stories.

2 What is the main gas found in the atmosphere? Where does it come from?

3 Explain the importance of the gases in the atmosphere for plants and animals. Draw a diagram to help explain this importance.

4 Find out about any geothermal area in your province or a province nearby. What creation stories are there about these places? Do the people make use of these places? How do they use them?

The shape of the earth

People have had many ideas about the shape of the earth. In earlier times some people thought that the earth was flat and that if a person went to the edge of it, that person would fall off. We now know that the earth is a sphere by the way that a ship seems to sink below the horizon when it sails away. We can also see the curved surface of the earth from an aeroplane high in the sky and satellites have also taken photographs of the earth from space showing that it is a sphere.

The fact that the earth is a sphere can be seen by the way that a ship appears to sink below the horizon.

The structure of the earth

The earth is made up of four different layers. Each layer is made of different materials. These layers are the **crust**, the **mantle**, the **outer core** and the **inner core**. If we were able to cut through the earth and remove part of it we would be able to see the different layers.

The crust

- The crust is the outer layer of the earth.
- Most of it is solid rock from 8 to 64 km thick, with a thin layer of soil.
- The crust is thickest under the continents and thinnest under the sea floor.
- The temperature varies from 20°C at the surface to 500°C near the mantle.

The mantle

- The mantle is the layer below the crust.
- It is about 2800 km thick.
- It is mostly hot molten rock that moves continually. This is what we see when a volcano erupts.

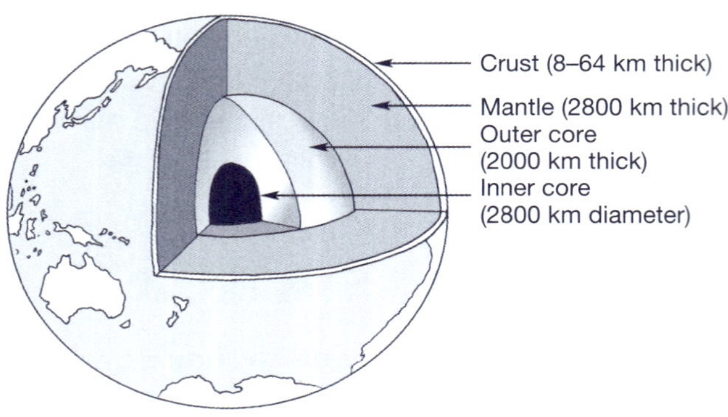

The earth with part of it cut away to show the different layers.

- The temperature varies from 500°C near the base of the crust to 2000°C near the outer core.

Outer core

- The outer core is just under the mantle.
- It is about 2000 km thick and contains molten iron and nickel, which is a silver-coloured metal.
- The temperature varies from 20 000°C to 3000°C.

Inner core

- The inner core is at the centre of the earth.
- It is 2800 km in diameter and made of iron and nickel.
- The temperature is about 4000°C. Although you might expect the inner core to be liquid because of the very high temperature, scientists think that it is solid because of the high pressure.

For you to try

1 Make a model of the earth with part of it cut away to show the four different layers. Another way would be to make a model showing a 'slice' of the earth, like a slice of cake. Use a different colour for each layer and label the layers.

2 Using the information in the table **Temperature of the earth's crust** (on p. 129), draw a graph to show how the temperature of the earth increases with depth below the surface. Put the depth on the horizontal axis and the temperature on the vertical axis. Choose a suitable scale. Remember to label each axis.

M A

- **a** Write a few sentences to explain what the graph tells us.
- **b** By how many degrees does the temperature rise each kilometre down into the earth's crust?
- **c** Use the graph to predict what the temperature would be at 500 metres, 3.5 km and 10 km.
- **d** Miners cannot work at temperatures above 45°C. At what depth will the temperature be 45°C?

3 Nobody has ever travelled inside the earth more than a few kilometres. Why is this?

4 From the information given on the structure of the earth, copy and complete the following table:

Structure of the earth

Layer	Thickness	Temperature	What is it made of?
Crust			
Mantle			
Outer core			
Inner core			

Rocks and soil

PD SS

People in Papua New Guinea have always made use of the rocks which they find in their surroundings. For example, special kinds of stones have been used for thousands of years. Stones which could be given a sharpened edge were used to make **axes** and **adzes**. These tools were used to cut down trees and to make canoes. In Manus, Milne Bay and West New Britain, a shiny black stone called **obsidian** was mined and traded with people from other places. This stone was used to make tools with a sharp edge and was highly valued. Obsidian was traded to the highlands 6000 years ago and was traded throughout New Ireland.

Other kinds of stones are used in cooking. Stones that can be heated in a fire for a long time can then be used to cook food in a ground oven, or **mumu**. This kind of cooking is popular throughout the Pacific Islands, especially for important occasions.

All of these special stones are really different kinds of rocks which are used because of their properties. Only rocks that are strong and can be sharpened are useful for making tools. Only rocks that do not break when heated are suitable for making a mumu.

Stones which could be sharpened were used to make axes and adzes.

Rocks that do not break when heated in a fire can be used for making a mumu.

Rocks

Igneous rocks

The first rocks formed on the earth when the molten rock or **magma** inside the earth began to cool down. When molten rock comes out of a volcano it is called **lava** and rocks are also formed when it cools. These rocks are called **igneous rocks** and are usually very hard. For example, basalt, granite and andesite are examples of igneous rock in Papua New Guinea. Basalt rocks can be seen near Sogeri, Lae, Mount Hagen and many other places in Papua New Guinea.

Basalt rocks

Obsidian is an igneous rock that looks like black glass.

Pumice is a very light igneous rock that floats on water.

Obsidian is a hard igneous rock that is formed when the lava from a volcano cools quickly in water. Obsidian looks like black glass and can be made into tools with a sharp edge. However, some igneous rocks are not hard. For example, the pumice that comes out of volcanoes has many holes in it that are formed by bubbles of gas in the rock. Pumice is so light that it can float on water.

Sedimentary rocks

When a river looks muddy, tiny pieces of rock are being carried in the water. When the river reaches the sea or lake the water slows down and the material being carried in the water falls to the sea floor and is called **sediment.** Over thousands of years, layers of sediments can build up and be squashed together to form new rocks called **sedimentary rocks**. Sandstone, mudstone, conglomerate and limestone are examples of sedimentary rock.

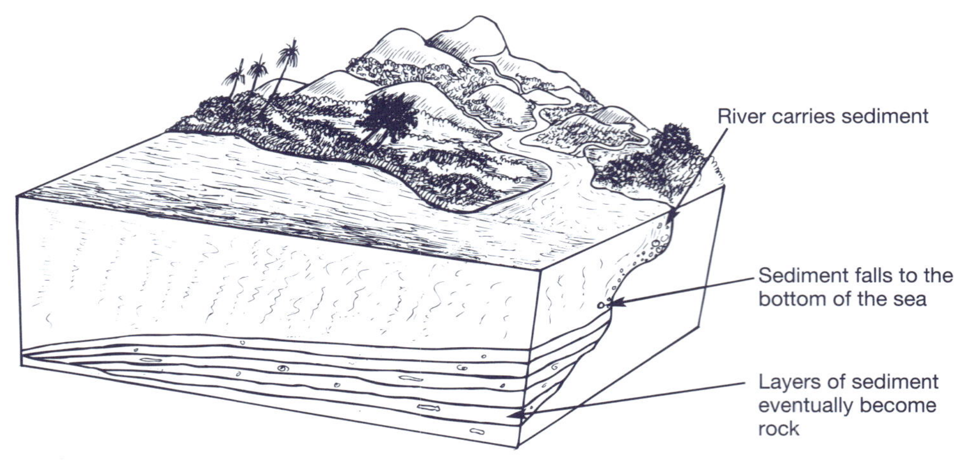

The formation of sedimentary rocks

Sedimentary rock often has layers. When sedimentary rocks are formed, the remains of plants and animals may be trapped in the layers. These remains become part of the rock and are called **fossils**.

Metamorphic rocks

Igneous and sedimentary rocks deep within the earth's crust are under high pressures and temperatures which usually make the rocks harder and make them look different. Rocks that are formed in this way are called **metamorphic rocks** because they have been changed. Slate is an example of metamorphic rock. The layers in slate are so well formed that it can be split into sheets.

Types of sedimentary rock

A fossil

Slate can be used to make roof and floor tiles.

Marble can be used to make floors and statues.

These split sheets can be used as tiles for roofs and floors in some countries. Slate is also used to make pool tables because a big sheet of slate is flat and smooth. Marble is another metamorphic rock that can be used to make floors and statues. The floor of the National Parliament in Port Moresby is made from marble.

Weathering and erosion

Rocks are slowly worn down by the sun, wind, rain and temperature. This process is called **weathering** and goes on all the time. For example, rocks can become very hot in the daytime and then get cold at night. These changes in temperature can make the rocks expand and contract and the surface of the rocks begin to break into smaller pieces.

In high mountains the temperature can fall below 0°C at night. Water in small cracks in the rock will freeze and turn to ice. Water expands when it freezes and the force is big enough to make the crack a little bit wider. Over a long period of time the crack will gradually become bigger and bigger and eventually a small piece of rock breaks off.

The roots of plants can also grow on rocks which can cause rocks to crack and break into smaller pieces.

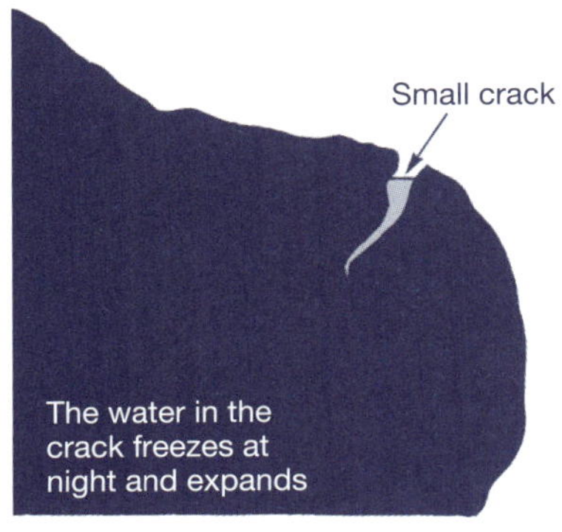

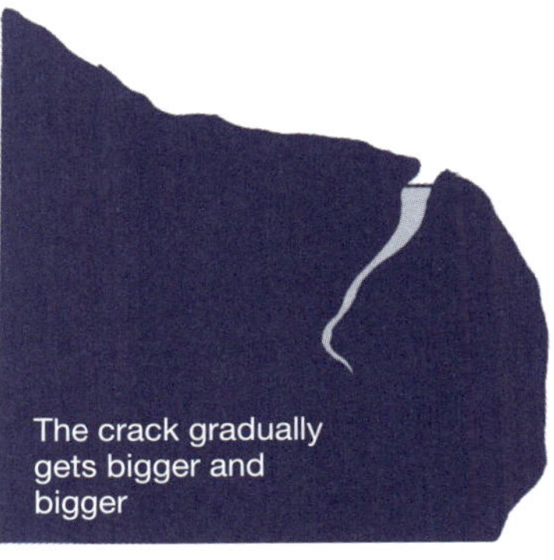

Water and temperature can cause weathering.

Plants growing on rocks can help to break them into smaller pieces.

Running water is the main cause of erosion in Papua New Guinea.

The small pieces of material that come from the weathering of rocks, and soil, may be carried away by water and wind. This process is called **erosion** and large amounts of material can be carried in this way. Running water is the main cause of erosion in Papua New Guinea.

When a river looks muddy, it is carrying tiny pieces of rock in **suspension**. This suspended material includes clay, silt and fine sand. These substances cannot settle because the water is always moving or **turbulent**.

However, the river slows down when it reaches the sea or a lake and the material carried settles down as a **sediment**. Layers of sediment can build up over thousands of years and become squashed and stick together to form new rocks called **sedimentary rocks**.

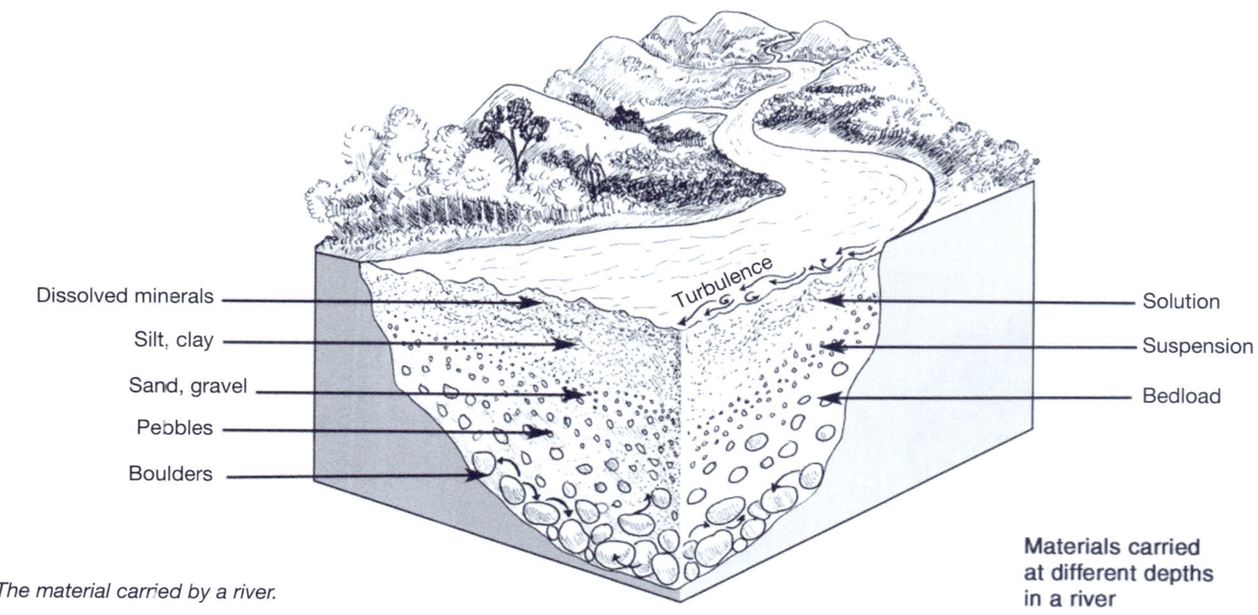

The material carried by a river.

When there is a flood there is a greater flow of water so that sand, pebbles or small stones and boulders roll or slide along the river bed.

Although weathering and erosion take place all the time, in some places the land is being pushed up to form mountains. This process is called **uplift** and is the way in which the mountains were formed in Papua New Guinea. These high areas of land are again weathered and eroded and the process continues.

The action of weathering and erosion also produces soils that can support plants. Plants provide the basic source of food for all animals. When plants and animals die some of the materials taken from the soil will be returned to the earth.

The rock cycle

All of the processes of rock formation, uplift, weathering and erosion that occur in the earth's crust are part of the **rock cycle.**

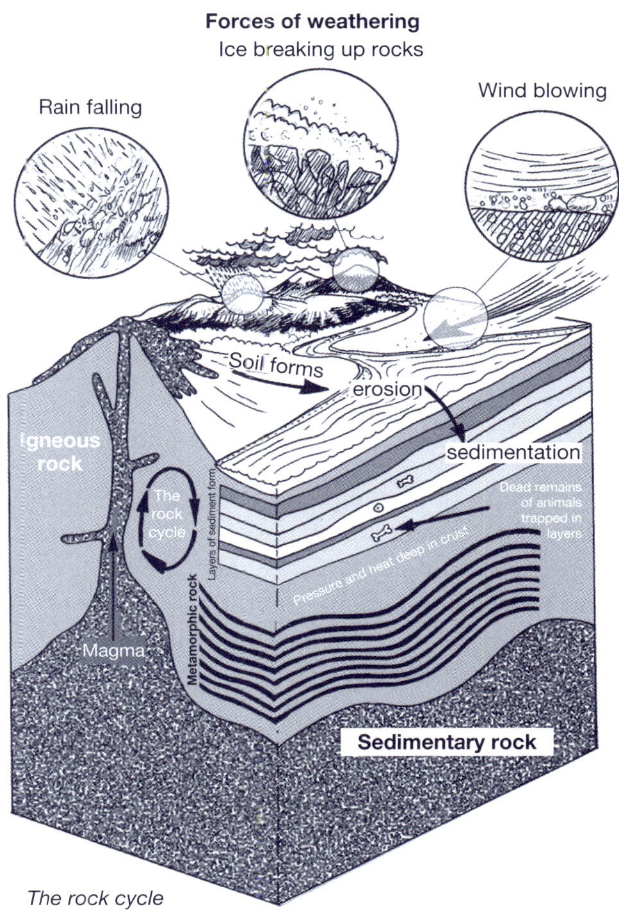

The rock cycle

For you to try

1 Make a visit to a mountainside, roadside, river or volcanic site to observe different types of rocks. Write a report of your visit and include a map and some diagrams to show what you found.

2 Make a poster to show the rock cycle.

3 **Investigation: Freezing and weathering**

   a Collect a plastic drink or cordial bottle and some water. You will also need to use the freezer section of a refrigerator.

   b Fill the bottle with water right to the top and screw the lid on tightly.

   c Put the bottle of water in the freezer and leave it until the next day.

   d Remove the bottle from the freezer and observe what has happened.

   e Discuss your observations with the other members of your group.

   f Explain what has happened to the water and the bottle.

Soils (PD)

Soils are formed from rocks that have been broken down over hundreds of years. Different types of soil come from different types of the rock and different sizes of particles. Soil consists of:

- tiny pieces of broken down rock
- decaying plants and animals or **humus**
- water and air trapped in spaces between the particles of soil
- living things such as earthworms and bacteria.

The amount of each of these things in the soil depends on the climate, the vegetation and the type of rock in the area. Since there are many different types of climate and different types of rock, soil types vary greatly from place to place.

Soil profiles

When we look at the face of a person from the side we see their outline or **profile.** We see the shape of their forehead, the nose, the mouth and the chin. Scientists who study soils are interested in the **soil profile**. They dig down through the soil in order to find out what the cross-section of the soil is like. Soil profiles can also be seen in a road cutting and sometimes along the bank of a river or on a hillside where there has been erosion.

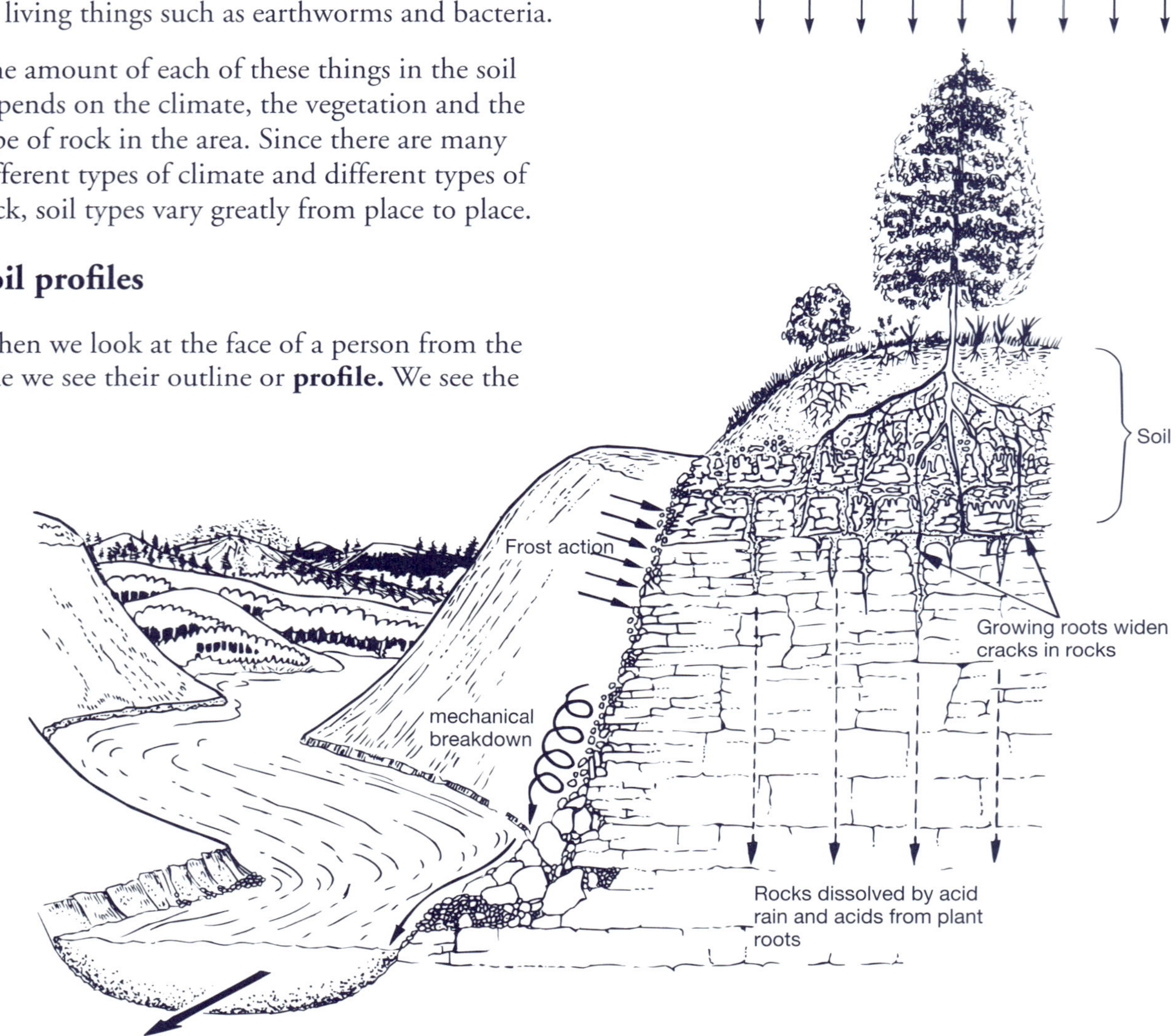

Soils are formed from broken down rocks.

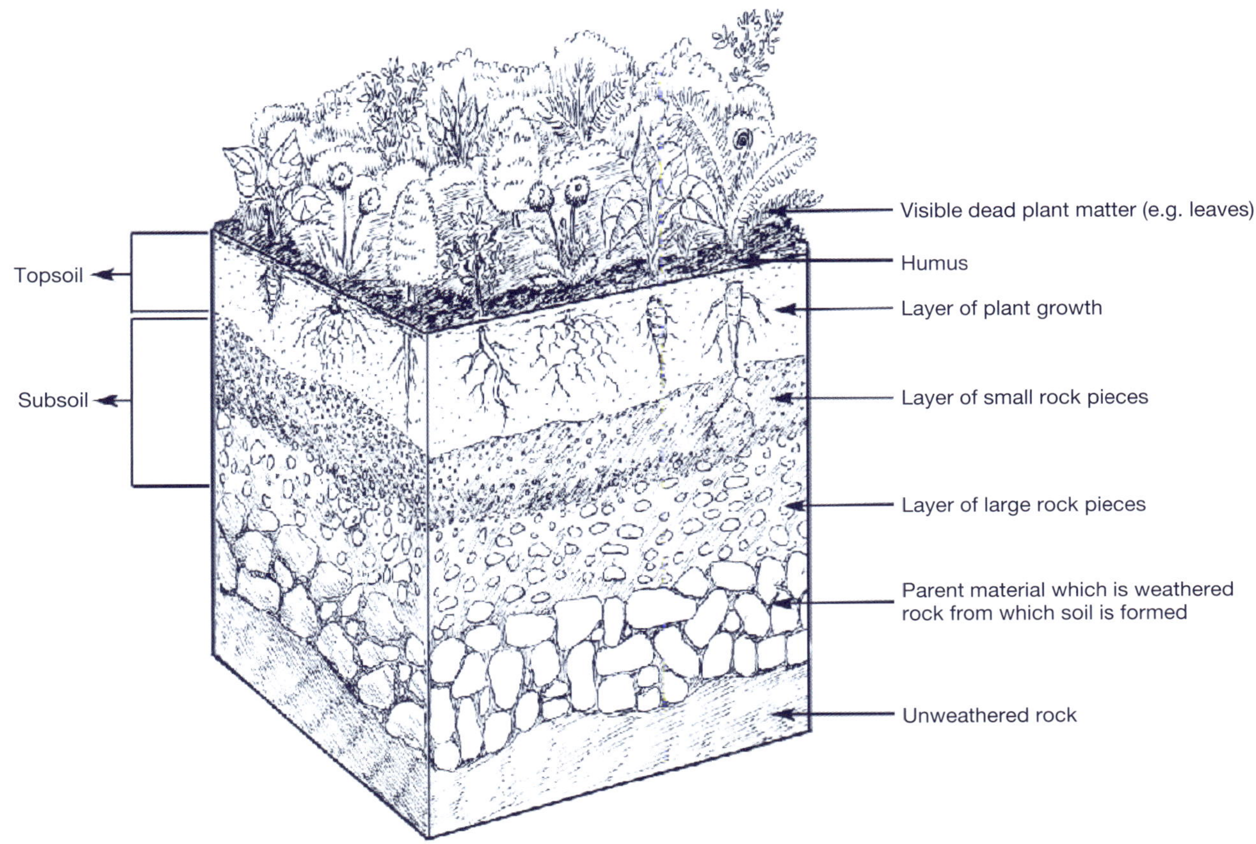

A soil profile showing the different layers

In most soils four distinct layers can be seen in the soil profile – **topsoil**, **subsoil**, **weathered bedrock** and **unweathered bedrock**. The weathered rock from which soil is made is also called its **parent material**.

Types of soil

When different rocks weather they form different types of soil. Different types of soils have different properties and different uses:

- A soil that contains a large number of sand particles is called a **sandy soil**. Sand particles are the biggest soil particles and so water will pass through easily. Sandy soils are well-drained but they may be too dry for many plants. Sandy soils are often found near the sea.
- A soil that contains a large number of clay particles is called a **clay soil**. Clay particles are the smallest soil particles and they stick together when wet, so clay soils are poorly drained and hold too much water for many plants.
- A soil that contains some sand, some clay and lots of humus is called a **loam soil**. Loam soil contains a mixture of large and small particles, mineral nutrients and organic matter and is good for growing crops.

The structure of a soil is important. If a soil has good structure, water can pass through it slowly. A soil with poor structure is more likely to be eroded.

When subsistence farmers make a new garden, the site is chosen carefully, as the kind of soil is very important for the crops to grow well.

Clay soils have very small particles that stick together and are not good for growing crops, but people have found another use for clay. When mixed with water, clay can be shaped into pots. After drying and firing, the clay pots can be used for cooking. They can also be traded with people who do not have clay soil in their area. This traditional skill is practised in many parts of Papua New Guinea. Very old pieces of pottery called **lapita pottery** are found throughout the Pacific Islands and show the possible trade and migration routes of people in the area.

Clay can be used to make pots.

For you to try

1 Investigation: Soil erosion

a Collect an old piece of corrugated iron or similar sheet, a fish tin with holes in the bottom to use as a sprinkler, and some grass or turf that you can dig up.

b Using soil, stones and sticks make a model of the side of a hill. Push the soil down well so that it is hard. You can put some stones on the surface and plant some small branches for trees.

c Using the tin with holes in the bottom, make it 'rain' on your 'hillside'. You can decide how heavy the 'rain' will be and how long the 'rain' will last.

d Observe what happens to the soil on your 'hillside'. Are there places where the erosion is greater and places where the erosion is less?

e How can you reduce soil erosion on your 'hillside'? Try your ideas to see if they work.

f Describe and explain your observations.

2 Investigation: Looking at soil

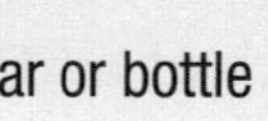

a Collect a tall glass or plastic jar or bottle and some soil.

b Fill the jar or bottle three-quarters full with water.

c Add soil to the jar or bottle but leave a space above the water.

d Put the lid on or put your hand over the end, shake the contents and allow them to settle.

e Observe what happens. You may have to wait for several days for everything to settle. It is best not to move the container during this time.

f Draw a diagram to show what happens. Label your diagram to show the size of the different particles and their size.

3 Visit a roadside cutting, hillside or river bank and make a labelled drawing of the soil profile. Try to identify the different layers and the type of topsoil.

4 Explain how it is possible for particles of igneous rock to become part of the soil in your garden, and to become sedimentary rock or metamorphic rock. Use a diagram of the rock cycle to help you and say how long these processes might take.

5 Project: make a rock collection

a Collect samples of rock from your area and other areas.

b Use the table below to identify the rocks and classify them as being igneous, sedimentary or metamorphic.

c Label the rocks to show where the rock came from, the date and name of the rock. Display your rocks for others to see.

d If you live in a limestone area you can look for and make a collection of fossils.

Characteristics of different types of rock

Igneous	Sedimentary	Metamorphic
Made of crystals—may be very small or large	Made of grains—may be cemented together	Crystals are lined up
Usually very hard and not easily broken	Not so hard, quite easy to break	Usually hard
No layering	Usually has layers	May be flaky, have layers or split easily
Colour from light to dark	May contain fossils	

Space exploration

For centuries people have looked up at the night sky and wondered about the moon and the stars and what was beyond the stars. Over many years people have made observations and kept records of events in the sky in order to understand what they saw. The people who do this are called astronomers. **Astronomy** is the scientific study of the movement of the sun, moon, planets and stars.

The things that people observe moving in the sky are also known as **heavenly bodies**. Because they move in a regular way, people learned how to use the movement of heavenly bodies. For example, sailors, fisherman and hunters use the stars for **navigation** or to find their direction. People also discovered that the movement of heavenly bodies could be used to make a **calendar** to measure time. They followed these movements and used them to count days, months, seasons and years. Farmers used the phases of the moon and the movement of the sun each year to know when to plant and when the rainy season would come. People have also used the movement of heavenly bodies to set the time for important events and trading expeditions.

Traditional calendars based on these regular cycles were made in Babylon and Egypt before 2500 BC. The Egyptians divided their farming year into three seasons. The year began when a certain star first appeared at a spot in the sky. When this happened it meant that the River Nile was about to flood. The flood left behind a rich layer of soil which was good for growing crops.

Much later the Mayan Indians of Central America used the movement of the sun to set the time for burning off their cornfields before planting each year. They also checked their calendar against measurements of the movement of the planet Venus.

The sun, moon and stars are all heavenly bodies.

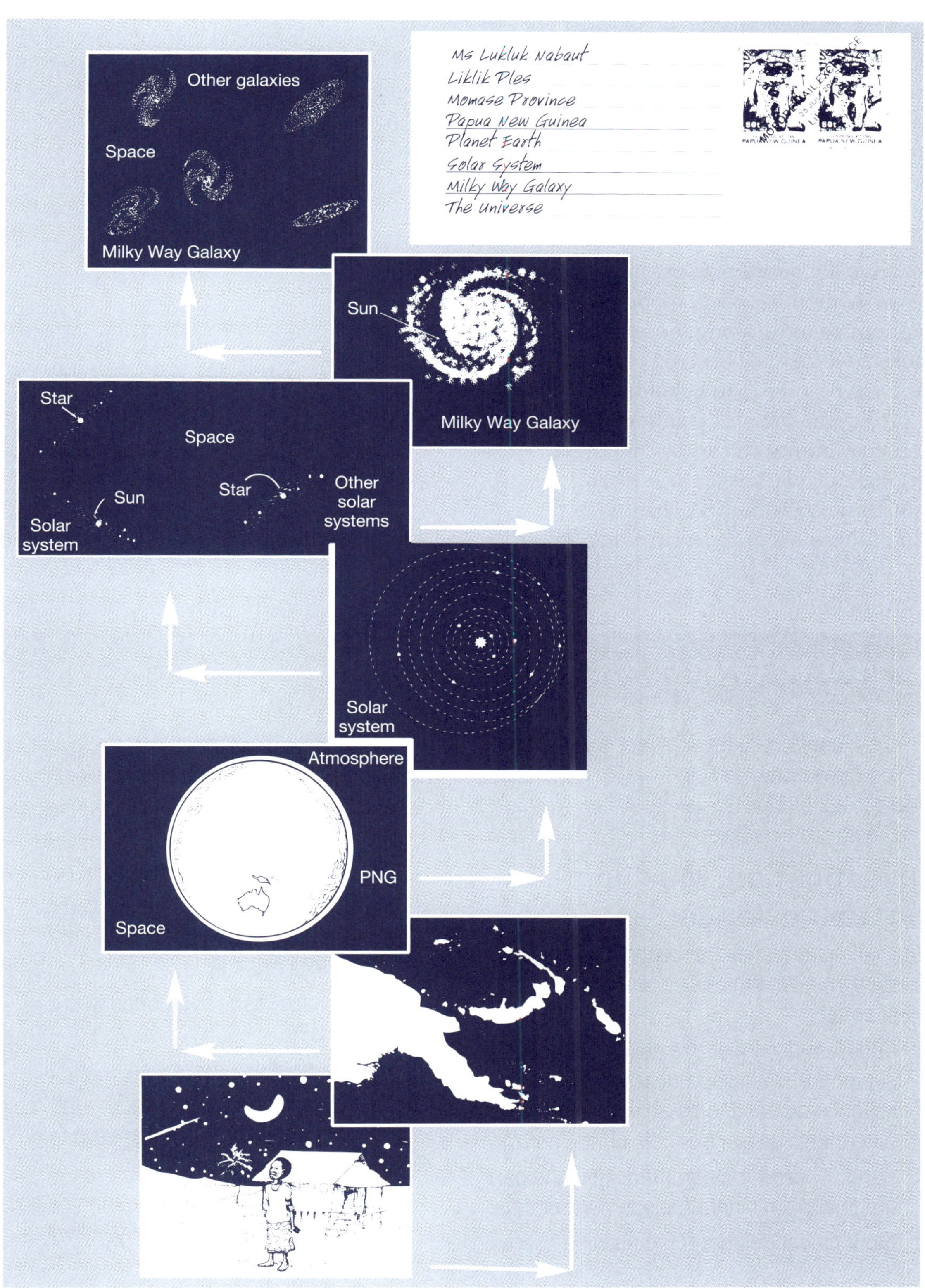

Our place in the universe

In Papua New Guinea, the sky is also used as a clock, calendar and compass. The movement of heavenly bodies can be used to arrange trading trips, feasts and ceremonies, and gardening activities. For example, the sun does not rise and set at exactly the same place on the horizon each day but rises and sets a little bit further north or south depending on the season. If you observe the position where the sun rises or sets over six months or a year you will notice that it appears to move along the horizon and then comes back again. This type of traditional calendar is called a **horizon calendar** and can be used to predict the time for planting or harvesting and other community activities. Other types of traditional calendar make use of changes in the seasons, especially rainfall, changes in the direction of the wind, and the time that flowers appear on some trees. These changes are **natural clocks** that cannot be changed by people and so they help to plan the calendar and set the time for different activities.

PD

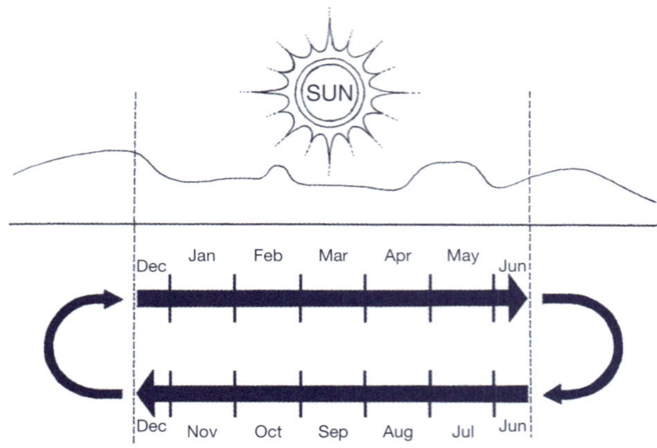

A horizon calendar

For you to try

1 Find out about traditional beliefs and stories in your area about the night sky and the stars. Write these stories or make a poster to share them with other groups.

2 **Investigation: Making a horizon calendar** A

You will need to make observations for at least six months but twelve months would be even better.

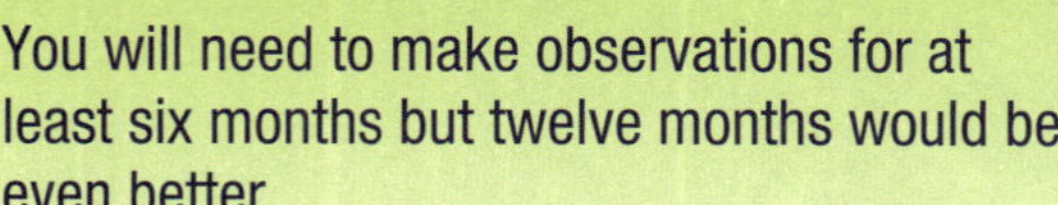

- **a** Choose a good place to observe the rising sun or the setting sun near your home or school. You need to be able to see clearly where the sun rises or sets on the horizon.
- **b** Carefully draw a diagram to show the part of the horizon where the sun rises or sets.
- **c** Mark on your diagram the position of any features such as hills, trees or buildings that you can see against the sky.
- **d** Observe the sun from this same position over many weeks and mark on your diagram where the sun rises or sets.
- **e** How long does it take before you can notice a change in the position of the sun as it rises or sets?
- **f** What happens over a period of six months?
- **g** What happens after twelve months?
- **h** When does the sun appear to stop moving along the horizon and then start to go back the other way?
- **i** Mark on your calendar the months and some important community activities.

j Use your horizon calendar to predict some important events, like school holidays for example.

k Show your horizon calendar to other people.

3 Make a traditional, school or community calendar to show the activities and events that occur in different seasons. Briefly explain how the activities and events are related to the way of life.

Science in the village: Traditional Trobriand knowledge  

Wawela is a village on the eastern side of Losuia in the Trobriand Islands of Milne Bay Province. Wawela is the centre of traditional astronomical knowledge in the area and the traditional calendar of the people has been worked out here for many years.

The people of the Trobriand Islands have names for many stars and groups of stars. The most important group of stars is called ***Uluwa***. Uluwa is the Trobriand name for the group of stars known by European astronomers as ***Pleiades*** or Seven Sisters. This group of stars is also important in the calendar of the Maori people of New Zealand and people all over Africa use the appearance of this group of stars to begin their farming each year. People all over the world recognise and use this group of stars more than any other.

Another important star is called ***Kibi***, which is a single star. The people of the Trobriand Islands understand the annual cycle of Kibi and Uluwa and use this information to set the time for preparing, planting and harvesting their yam crops which are very important in the Trobriand Islands.

Other important stars are ***Munikaiwau*** and ***Mitubuyubuyai***. Munikaiwau is the brightest star in the sky apart from the planet Venus, which is not a star. Munikaiwau is called ***Sirius*** by European astronomers. Mitubuyubuyai or Betelgeuse (pronounced 'betel juice') is a big red star.

People in the Trobriands also give the names of familiar objects to other groups of stars. These include ***Sinata*** the hair comb, ***Lakum*** the crab, ***Osukwalu*** the fish net and ***Kekyiadiga***, which means 'precision' or 'accuracy', that has three stars in a line. In European astronomy, Kekyiadiga is well

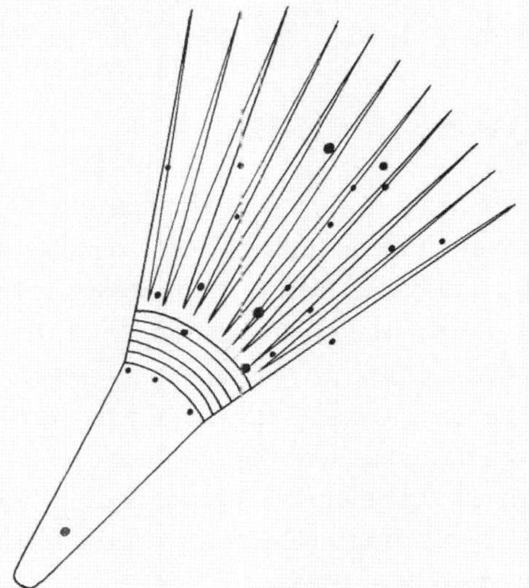

The group of stars given the name **Sinata** *by people in the Trobriand Islands because it looks like a hair comb.*

known as Orion's Belt. The Greeks could see the shape of a hunter in the pattern of the stars and they called this hunter ***Orion***.

Another important set of stars that can only be seen by people in the southern hemisphere is the Southern Cross. The Southern Cross is the set of stars shown on the PNG national flag.

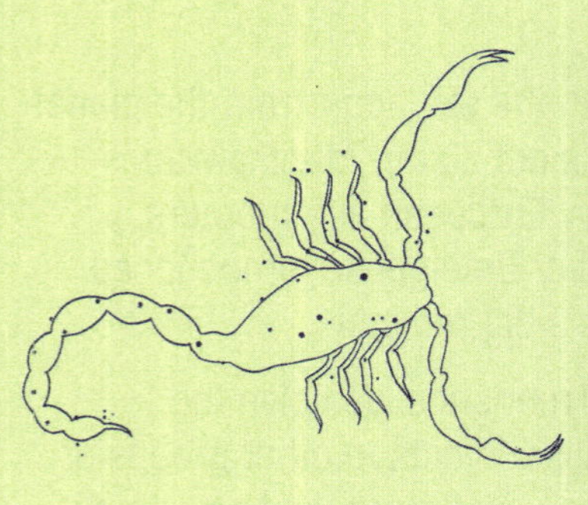

The ancient Greeks called the same set of stars **Scorpio** *because they thought it looked like a scorpion.*

The ancient Greeks thought that this pattern of stars looked like a hunter so they called it **Orion**.

Using a star chart

A map can be used to represent the features found on the earth such as mountains and rivers, towns and villages, roads and bridges. A map of the sea showing reefs and islands and the depth of the water is called a **chart**. In the same way a **star chart** can be used to represent the objects found in the sky. A star chart like the one below is really a map of the sky. The observer reads the chart as if he or she is standing in the centre of the circle, and the chart is the sky above.

At certain times of the year, Kekyiadiga or Orion's Belt appears almost directly overhead. This is the same every year. This star chart shows these times and the table will help you to look for stars.

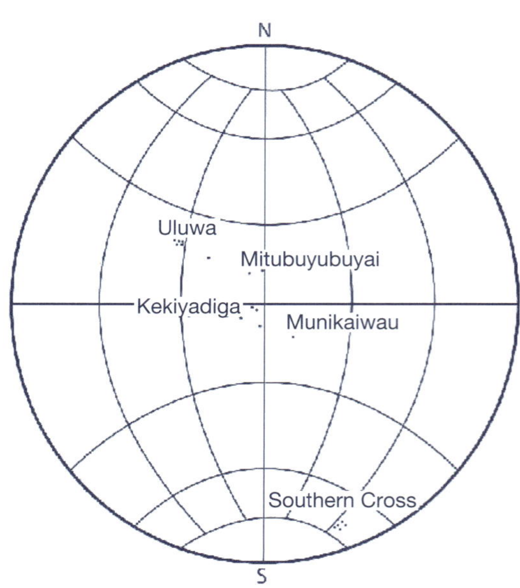

A star chart

Looking for stars

Name	Where do you look?	Description
Munikaiwau or Sirius	Look to the south	Brightest star to the left of Orion's Belt
Mitubuyubuyai or Betelgeuse	Look to the north	Big red star slightly to the north of Orion's Belt
Uluwa or Pleiades	Look to the north, to the left of Orion's Belt	Uluwa twinkles and looks like six stars close together (several hundred stars when viewed through a telescope)
The Southern Cross	Look to the south, low in the sky, to the left	Four stars in the shape of a cross, and two stars known as pointers

The sky appears to rotate once each day as the earth rotates. The sky also appears to make one complete rotation each year.

If the sky is observed every night at the same time, it can be seen that the sky seems to turn a little to the west each night. It is this annual movement that helps people to use the stars as a calendar.

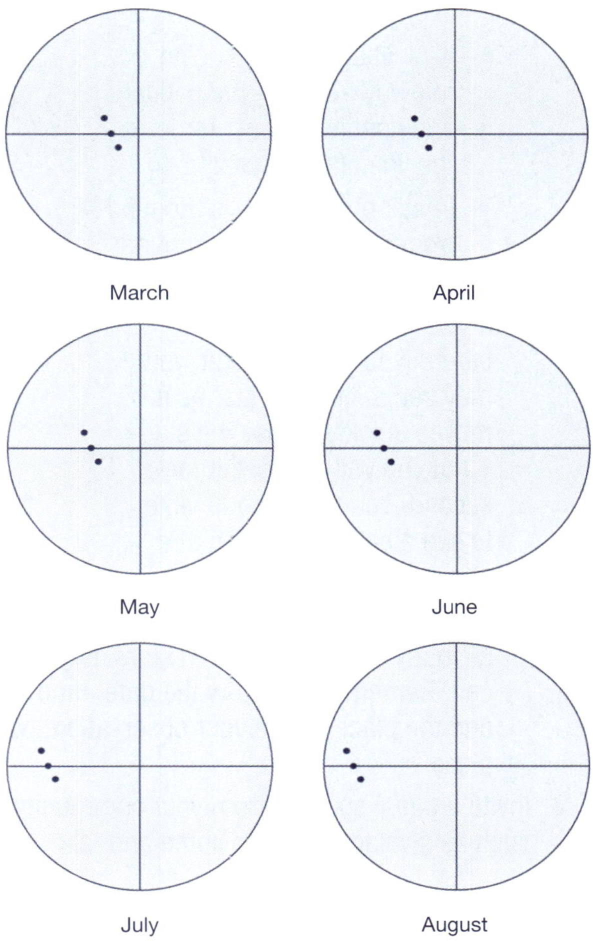

The positions of Kekiyadiga, or Orion's Belt, over a period of six months.

For you to try

1 Investigation: Looking at the night sky PD

a Collect a copy of the **star chart** and the table **Looking for stars** (above) and a torch.

b Go outside on a clear, dark night. Look up into the sky. What can you see?

c Using the **star chart** and the table **Looking for stars**, identify different stars and groups of stars.

d Once you have found the Southern Cross, follow these steps to find south more accurately:

- Take the long axis of the Cross and extend it by imagining that the line continues into space beyond the Southern Cross.

- Now imagine a line going midway between the pointers and continuing until it meets the line of the Cross.
- Imagine that a line is dropped from this point to the ground – this is the south.

e If you are lucky and look carefully for long enough, you may see a shooting star as it moves quickly across the sky – but you will only see it for a second. You may also be able to see a satellite moving at a steady speed high in the sky.

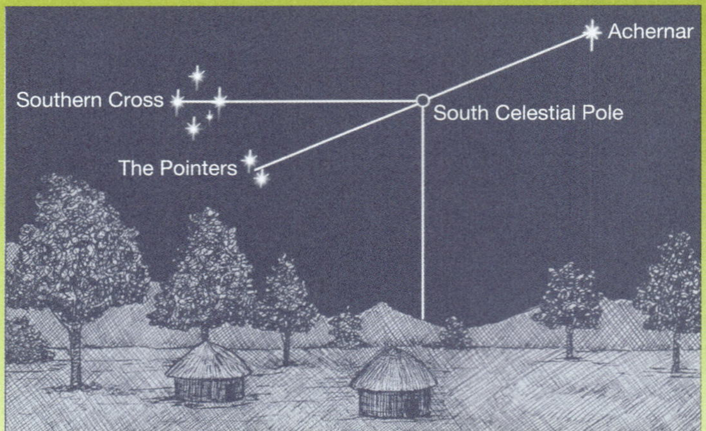

Using the Southern Cross to find south

f Draw a star chart or map of the sky. Label as many stars or groups of stars as you can. Remember to show the date, time and the place where your observation took place.

2 Invite a guest speaker from your community, such as a village elder, to come and talk about the way in which they use star patterns and the moon in their daily activities.

3 Ask older people in your community if they know of any myths or legends about the stars. Share these stories with the rest of the class.

4 Find out about traditional beliefs that explain how the earth and stars were formed.

The moon

Apart from the sun, the moon seems to be the largest object in the sky. In earlier times people looked at the moon and wondered what it was, what it was made of and whether there was life on it.

Because the moon shines at night, changes shape and seems to affect many things it has always been important in farming, religion and the way that people live.

Today we know a lot about the moon. Since 1966 scientists have sent spacecraft to visit the moon and take photographs. On 20 July 1969 the first men landed on the moon. Since

The moon shown during the last quarter.

Astronauts walking on the moon.

then others have visited the moon, carried out experiments and brought back samples of moon rock.

The characteristics of the moon

The moon is the earth's nearest neighbour in space and is about one quarter of the size of the earth.

The earth orbits the earth once every 28 days, so it takes 28 days for the moon to travel around the earth. The moon also rotates on its own axis once every 28 days, so the moon always has the same side facing the earth. The far side of the moon had never been seen until photographs were taken by the first space probe which visited the moon in 1959.

The moon does not give out any of its own light. We see the moon by sunlight reflected off its surface.

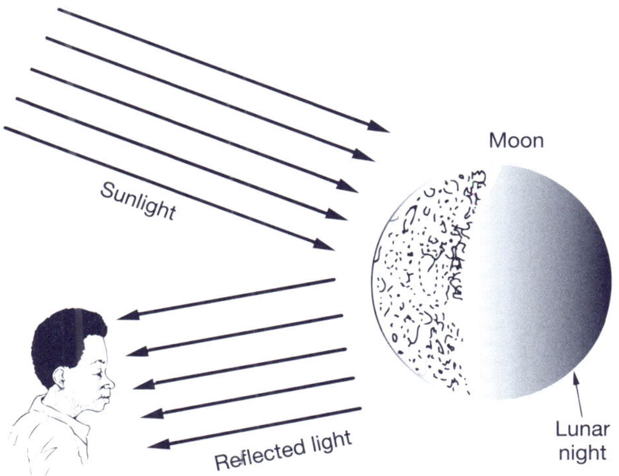

We see the moon by the sunlight it reflects.

The phases of the moon

Every twenty-eight days, the moon goes through a cycle of light and dark shapes. The different shapes of the moon seen from the earth are called the **phases** of the moon. These different shapes are not caused by the earth's shadow on the moon but by the moon's daytime and night-time.

Even though the moon is a long way from the earth, the moon can change things on earth. For example, people who live near the sea know that the sea rises and falls at different times of the day and night. This is called **high tide** and **low tide** and is caused by the pull of the moon on the earth.

The highest tide, called the **spring tide**, usually occurs at full moon and new moon. The lowest tide, called **neap tide**, occurs during the first quarter and last quarter.

Tide charts are printed in special books and are also found in newspapers. The chart gives the time of high and low tide, and the height of the water in metres for different places.

The moon is used to set the time of Easter each year. Easter is always on the first Sunday after the first full moon after the equinox. The equinox is

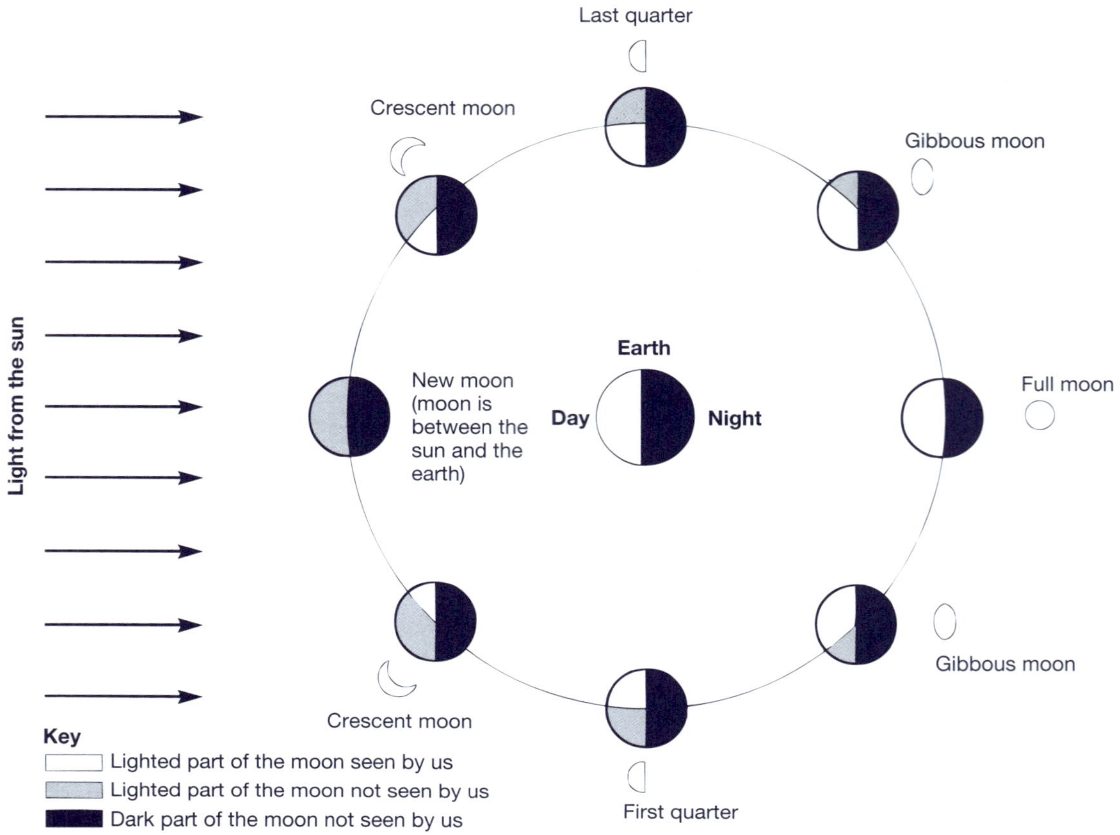

Last quarter

Crescent moon

Gibbous moon

Light from the sun

Earth

New moon (moon is between the sun and the earth)

Day

Night

Full moon

Crescent moon

Gibbous moon

First quarter

Key

Lighted part of the moon seen by us

Lighted part of the moon not seen by us

Dark part of the moon not seen by us

The phases of the moon

the time when the length of the day and night are the same, about March 21. For this reason, Easter is sometimes in March and sometimes in April, although Christmas is always on the same day.

Many people believe that the behaviour of living things is affected by the moon. For this reason, gardeners, hunters and fishermen often arrange their activities according to the phases of the moon. The word 'lunatic' means mad and comes from the Latin word *lunar* which means moon. The behaviour of mad people sometimes seems to change with the phases of the moon.

The highest tide occurs at full moon and new moon. The lowest tide occurs at first quarter and last quarter.

For you to try

1 Investigation: Looking at the moon

a Observe the phases of the moon over one month.

b Record your results by drawing a series of diagrams on a chart to show the way that the shape of the moon changes. Use the example below to help you.

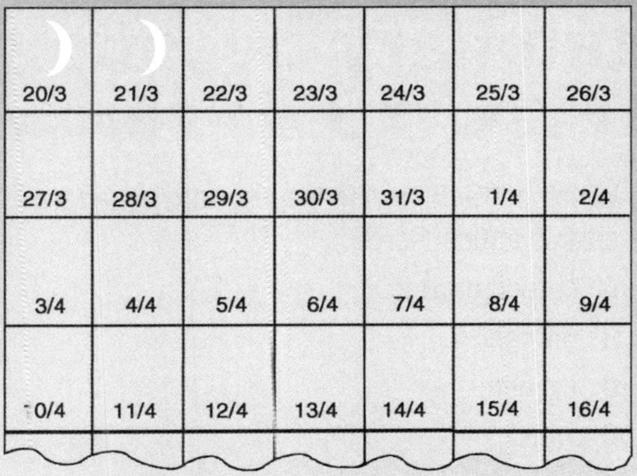

c Describe your observations.

20/3	21/3	22/3	23/3	24/3	25/3	26/3
27/3	28/3	29/3	30/3	31/3	1/4	2/4
3/4	4/4	5/4	6/4	7/4	8/4	9/4
10/4	11/4	12/4	13/4	14/4	15/4	16/4

2 Draw a traditional calendar that includes seasonal changes and the phases of the moon.

3 Explain how the phases of the moon affect your daily life or the life of people in your area.

4 Look up the tide chart in a newspaper. Choose one of the places mentioned in the tide chart.

a How many high and low tides were there for the day that is described?

b What was the time of high tide and low tide?

c How high was the tide?

d Compare this with another place that is mentioned in the chart.

e Write a brief summary to describe the tides in the two different places on that day.

Summary questions

1 Which of the following is most common in the earth's atmosphere?

A carbon dioxide

B nitrogen

C oxygen

D water

2 Which of the following gases is needed by most living things?

A carbon dioxide

B nitrogen

C oxygen

D water

3 Which of the following best describes the layers of the earth in order starting from the surface?

A core, mantle, crust

B crust, core, mantle

C mantle, crust, core

D crust, mantle, core

4 Which of the following best describes the changes that take place in the rock cycle?

A weathering → erosion → sedimentation

B erosion → weathering → sedimentation

C sedimentation → erosion → weathering

D weathering → sedimentation → erosion

5 Which of the following best describes the changes that take place during the formation of soil?

A topsoil → subsoil → bedrock

B bedrock → subsoil → topsoil

C bedrock → topsoil → subsoil

D subsoil → bedrock → topsoil

6 Which of the following best explains why people in traditional societies were able to use the stars and the moon to find their way or navigate and to measure time?

A The stars and the moon form patterns in the sky that change regularly.

B The stars and the moon can always be seen clearly every night.

C The stars and the moon always have a regular cycle of 28 days.

D The sky can be used as a clock, calendar and compass.

7 The passage below is a summary of the main ideas of this chapter. Copy and complete the passage in your book. Using the words in the list, find the words that are missing. You can use each word only once.

L

cycle	**erosion**	**formed**
ideas	**layers**	**navigate**
phases	**rock**	**sediment**
sea	**soil**	**stars**
water		

There are different ____________ about the way in which the earth was formed. The earth has three main ____________: the crust, mantle and core. Rock can be ___________ in different ways. For example, rocks can be formed from volcanoes and from the sediments that are carried in water and settle on the bottom of the ___________. The materials that make up rock can move in a ___________.

Rock can be slowly worn away by water and wind and the small particles can be carried by ___________ and settle down as sediments. Over thousands of years the ____________ can form more rock. Soil is formed from small particles of __________ and the living things that have died and rotted. __________ can be carried away by water and wind. This process is called soil __________.

__________ form a pattern in the sky that changes regularly with the seasons. The moon has different __________ that change in a cycle every 28 days. People use the stars and the moon to find their way or ____________ and to measure time, for example, when setting the time for important events, when planting and harvesting and when planning trading expeditions.

Student assessment templates

Sample 1: Group assessment template

Group assessment template

Date: ______________________________

Names of group members:

____

____

Name of group: ______________________________

Activity:

____

____

____

____

What things did your group do really well?

____

____

____

____

What things does your group need to improve?

____

____

____

____

What are you going to do to change the way your group works?

____

____

Sample 2: Term self-assessment template

My term report

Name:

I rate my working habits this term as:

I am a quiet worker	/10
I am a neat worker	/10
I finish my work on time	/10
I am able to work by myself	/10
I work well in groups	/10

What I did well this term

What I plan to improve on next term

What I would like to see changed next term

What I really liked doing this term

What I didn't like doing this term

Science in the community

A simple game about simple machines

To play the game you will need:

- one dice (you can make your own if you do not have one)
- one different marker for each player

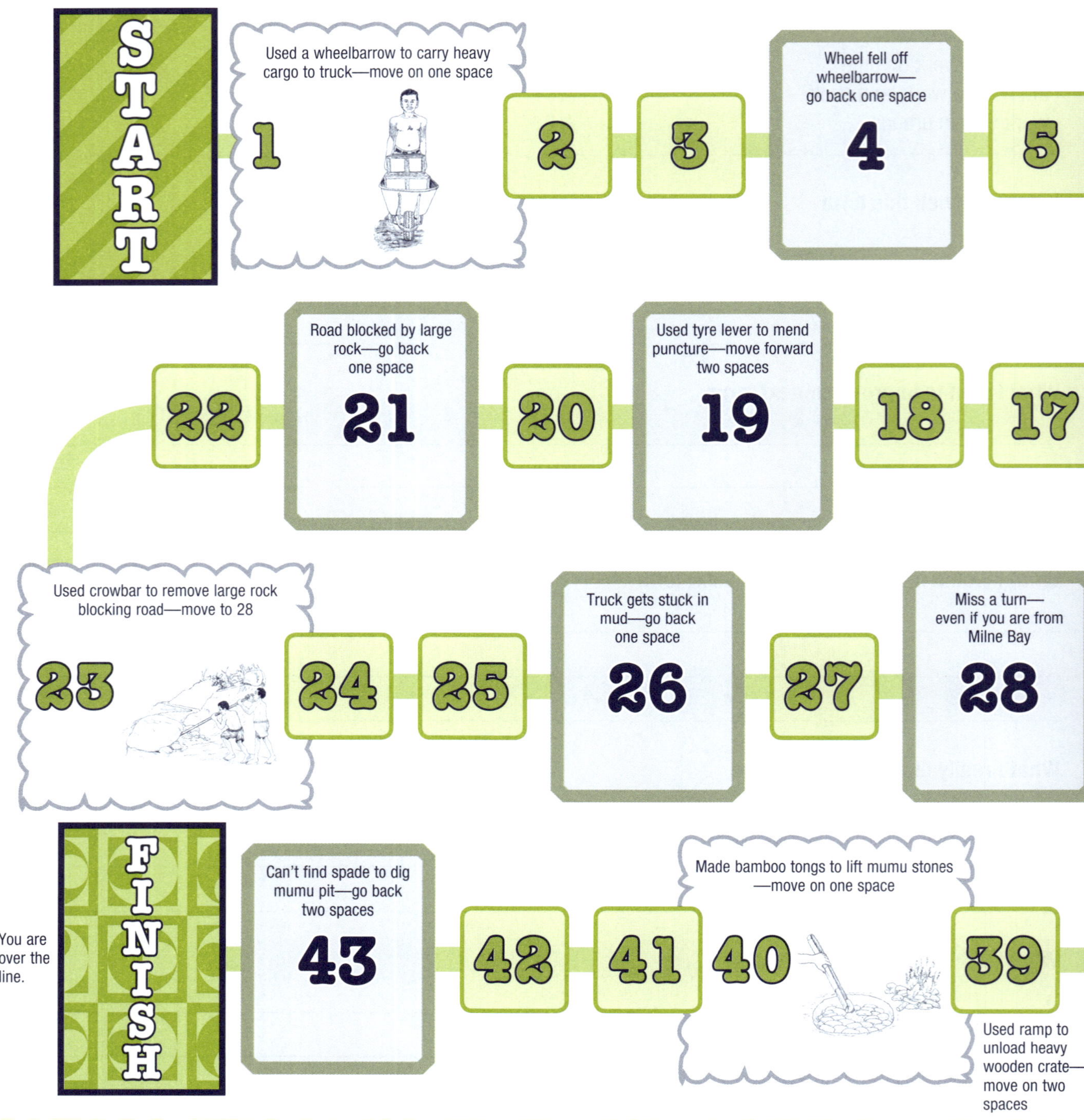

The first player throws the dice. The number on the dice is the number of squares that the player moves.

Each player then takes a turn.

If you land on a square with writing you must follow the instruction.

The winners are the ones who can use simple machines to help them in the community.

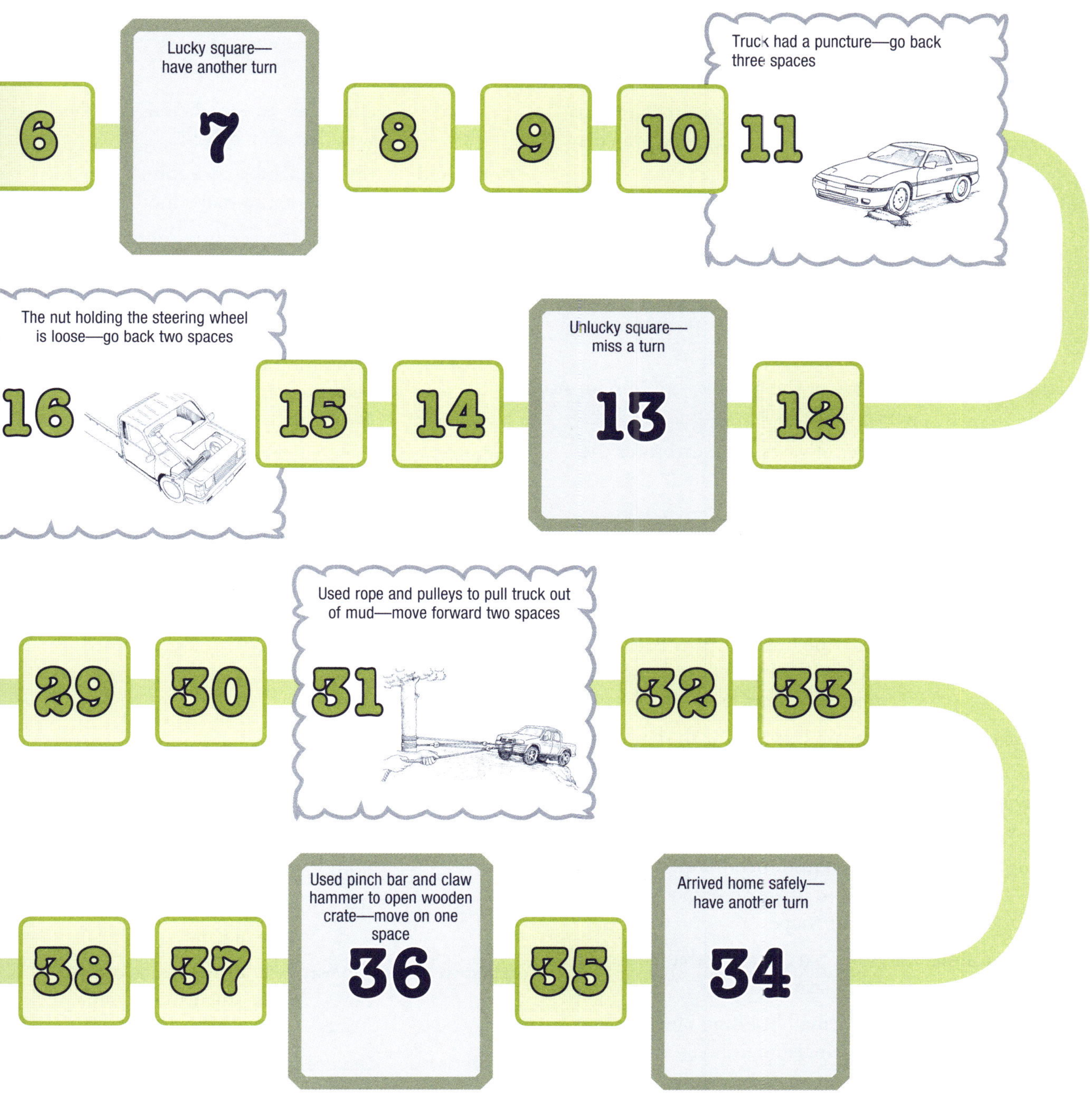

Glossary

adaptation the way a living thing has changed to make it more suitable for its way of life

algae simple green plants that usually live in water or damp places, e.g. seaweed

archaeology the study of the past and the way that people lived by looking at the things left behind, like pottery, tools, etc.

astronomy the study of the planets, sun, moon, stars and the universe

atmosphere the layer of gases that surrounds the earth

axle a rod that is connected to a wheel and which allows the wheel to turn

biology the study of living things like plants and animals

bush fallow allowing the bush to grow back in an old garden site so that the soil becomes more fertile

calibration to mark an instrument so that it can be used to measure accurately

carnivore an animal that eats other animals, e.g. spider

casting pouring liquid metal or plastic into a mould to make an object with a particular shape

cell membrane the outer part of a cell that contains the cytoplasm

cells the very small building blocks from which plants and animals are made

chemistry the study of substances like chemicals and medicines and the way that they behave

chlorophyll the green substance found in plants that is used to make food

classification putting things into groups according to their similarities and differences

compound a substance that is made by joining one or two elements together

conclusion something that you decide is true because you now know that other things are true

condensation the change of state from gas to liquid

conifers plants that have exposed seeds in cones, e.g. klinkii pine

consumer an animal that eats food

core the centre of the earth; there is an inner core and an outer core

creation stories a way of explaining how something was made

crust the outer layer of the earth

cytoplasm the liquid surrounding the nucleus of a cell that is contained inside the cell membrane

decantation separating an insoluble solid and a liquid by carefully pouring

development the way in which living things grow and become bigger

diffusion the way that a gas spreads out (so that it can be smelt in different places)

distillation the process in which a liquid evaporates to a gas and then condenses back to a liquid; distillation can be used to separate substances

ecology the study of the relationships and interactions between living things and their surroundings

elastic something which can stretch when a force is used and then return to its original shape and size

energy energy is the ability to do work; there are different types of energy

element a simple substance, e.g. copper, gold, oxygen

erosion small pieces of weathered rock and soil that are carried away by water and wind

evaporation the change of state from liquid to gas; also a method to separate a soluble solid from a solution

fern a green plant that has no flowers but reproduces by producing spores on the underside of the fronds

filter feeder an animal that obtains food by filtering the water in which it lives

filter paper a special kind of paper that is used to separate an insoluble solid from a liquid

filtration separating an insoluble solid and a liquid by pouring through a fine net which allows the liquid to pass through but catches the solid

flowering plants plants that have enclosed seeds formed within a fruit that develops from a flower

food chain a feeding relationship in which one living thing feeds on another and which always begins with a plant

food web a feeding relationship which is made up of a number of food chains; see also **food chain**

force a push or a pull; usually makes something move or change direction

fossil the remains of plants or animals that lived thousands of years ago

fossil fuel coal, oil and natural gas

freezing the change of state from liquid to solid; see also **solidification**

friction a force that slows things down or stops them moving; friction also allows things to grip so that they can move

fuel a substance that can be burned to give energy

fulcrum a fixed point or pivot that is used to make a lever work

fungi plants that have no chlorophyll but live on the dead remains of other plants and animals

gas matter that has no fixed volume and no fixed shape; see also **diffusion**

gears wheels with teeth around their edges that fit into the teeth of another wheel or into the holes of a chain

geology the study of rocks and the earth

geothermal heat from the rocks in the earth

gravity the force that pulls objects down

health the study of the reasons why we get sick and how we can get better

heavenly bodies the sun, moon, planets and stars

herbivore an animal that eats plants, e.g. butterfly

horizon calendar a traditional calendar that uses changes in the position of the sun as it rises and sets each day

humus decaying plants and animals

hydro-electricity electricity made by generators driven by running water

igneous rock rock formed from magma that has cooled and solidified at the earth's surface (volcanic rock) or deep within the earth's surface (plutonic rock)

incline a slope

inelastic something that does not return to its original shape when the force is removed

inference a type of conclusion and possible explanation

insoluble something that does not dissolve

invertebrate an animal that does not have a backbone

lapita pottery old pieces of pottery that show the possible trade and migration routes of Pacific Islands people

lever a simple machine that can be used to make work easier; a lever is usually a strong stiff bar or rod

liquid matter that has a fixed volume but no fixed shape

liverwort a simple green land plant that lives in cool, damp places

machine something that helps to make work easier

magma molten rock inside the earth

magnet a piece of iron or steel that attracts iron or steel towards it

mass the quantity of substance or amount of material in something

matter any material or substance that has mass and takes up space

melting the change of state from solid to liquid

metal solids that are usually hard and can be flattened into sheets and stretched into wires

metamorphic rock rock that has been changed because of high temperature and pressure; it is harder and looks different

meteorology the study of the weather

mixture something that contains at least two separate substances

molten rock or metal that has been heated to a very high temperature and has become a thick sticky liquid

moss a simple green land plant that lives in cool, damp places

navigation the process or skill with which people work out their position and direction when travelling

neap tide the lowest tide which occurs during the first quarter and last quarter of the moon

nucleus the part of a cell that controls the cell

obsidian a hard, shiny black stone that can be used to make sharp tools

omnivore an animal that eats both plants and animals, e.g. pig

organ a part of the body that is made up of two or more different kinds of tissue that work together to carry out a particular job, e.g. the heart

organism an animal or a plant; any living thing

pendulum something that swings backwards and forwards regularly

phases different shapes of the moon seen from the earth

photosynthesis the process in which plants make their own food using sunlight

physics the study of forces and movement, energy and matter, tools and machines

pivot see **fulcrum**

plane a surface

precipitation falling rain (also hail and snow)

producer a plant

properties the way in which a substance or object behaves

pulley a simple machine that consists of a rope, chain or belt stretched over the rim of a wheel

pure one substance that exists by itself

purification	to make something pure by removing unwanted substances
renewable	something that can be replaced
resource	something that can be used to meet the needs of people
respiration	the process in which plants and animals get energy from food
response	the way in which living things react to a stimulus
rock cycle	the processes of rock formation, uplift, weathering and erosion that occur in the earth's crust
scavenger	an animal that feeds on dead organisms
science	a way of finding out and understanding the world around us
scientists	people who work in science
sediment	small particles of solid that settle at the bottom of a liquid
sedimentary rock	rock that is formed from layers of sediments that have been compressed or squashed
soil profile	a cross-section of the soil showing the different layers
solar	from the sun
solar still	a way of obtaining fresh water from salty water by using distillation and energy from the sun
solid	matter that has a fixed volume and a fixed shape
solidification	the change of state from liquid to solid; see also **freezing**
soluble	able to dissolve
solute	a substance that dissolves in a liquid
solution	a mixture in which one substance is dissolved in another
solvent	a liquid that can dissolve things, e.g. water
spring tide	the highest tide which usually occurs at full moon and new moon.
state	the condition of matter, whether it is a solid, liquid or gas
substance	any material from which things are made
suspension	a mixture of small particles of insoluble solid and a liquid; see also **sediment**
switchback	a road or path that winds backwards and forwards across a hill side in a zigzag pattern
system	a group of organs and tissues that work together to carry out a particular job, e.g. circulatory system
thread	an inclined plane that turns round and round a screw in a spiral pattern
tissue	a group of cells that are similar to each other, e.g. skin tissue
traditional knowledge	the understanding of processes and materials that people use to live successfully
trial and error	trying out different ideas to see if they work
turbulent	water or air that is constantly moving and changing direction
uplift	movements in the earth which push up the rocks to form mountains
vertebrate	an animal that has a backbone
water cycle	the movement of water in the environment which includes precipitation, evaporation and condensation
weathering	the slow breaking down of rock into smaller pieces
wedge	something which is thick at one end and thin at the other, e.g. the head of an axe
weight	the downward pull of the earth's gravity on the mass of an object

Acknowledgments

The author and publisher wish to thank the following copyright holders for granting permission to reproduce their material. Sources are as follows:

Andrew Alden, geology.about.com, copyright 2006, reproduced under educational fair use, p. 135 (left); Photographs courtesy of Crawford House, pp. 5, 22, 44 (left), 78, 128, 142; The Australian Bureau of Meteorology, p. 92; FCX_Freeport-McMoRan Copper & Gold Inc., pp. 54, 55; Fotolia, pp. 27 (right), 29, 40, 98 (right), 117 (right), 135 (bottom right), 138 (right); iStock Photography, pp. 24, 27 (left), 44 (left), 71 (top right), 74, 75, 76, 77, 96, 129, 136; Permission by Orica Australia Pty Ltd., p. 87; Photolibrary, p. 107 (right); Irene Sawczak, pp. 42, 51 (bottom left), 63 (right), 95, 111; Smithsonian Institution, image #EMP21.009 from the Science Service Historical Images Collection, courtesy General Electric, p. 97 (left); Max Waugh, www.maxwaugh.com, p. 51 (bottom right).

Every effort has been made to trace the original source of copyright material contained in this book. The publisher would be pleased to hear from copyright holders to rectify any errors or omissions.